FRITJOF CAPRA AND THE SYSTEMS VIEW OF LIFE

Great Minds Series by Peter Fritz Walter

Krishnamurti and the Psychological Revolution

Creative Genius: Four-Quadrant Creativity in the Lives and Works of Leonardo da Vinci, Wilhelm Reich, Albert Einstein, Svjatoslav Richter and Keith Jarrett

Fritjof Capra and the Systems View of Life

Françoise Dolto and Child Psychoanalysis

Edward de Bono and the Mechanism of Mind

Joseph Murphy and the Power of Your Subconscious Mind

Joseph Campbell and the Lunar Bull

Terence McKenna and Ethnopharmacology

Charles W. Leadbeater and the Inner Life

Antonio Villoldo and Healing the Luminous Body

Wilhelm Reich and the Function of the Orgasm

FRITJOF CAPRA AND THE SYSTEMS VIEW OF LIFE

Short Biography, Book Reviews, Quotes, and Comments (Great Minds Series, Vol. 3)

by Peter Fritz Walter

Published by Sirius-C Media Galaxy LLC

113 Barksdale Professional Center, Newark, Delaware, USA

©2015 Peter Fritz Walter. Some rights reserved.

2017 Revised, Updated and Reformatted Edition.

Creative Commons Attribution 4.0 International License

This publication may be distributed, used for an adaptation or for derivative works, also for commercial purposes, as long as the rights of the author are attributed. The attribution must be given to the best of the user's ability with the information available. Third party licenses or copyright of quoted resources are untouched by this license and remain under their own license.

The moral right of the author has been asserted

Set in Palatino

Designed by Peter Fritz Walter

ISBN 978-1-514757-18-5

Publishing Categories
Science / System Theory

Publisher Contact Information
publisher@sirius-c-publishing.com
http://sirius-c-publishing.com

Author Contact Information
pfw@peterfritzwalter.com

About Dr. Peter Fritz Walter
http://peterfritzwalter.com

About the Author

Parallel to an international law career in Germany, Switzerland and the United States, Dr. Peter Fritz Walter (Pierre) focused upon fine art, cookery, astrology, musical performance, social sciences and humanities.

He started writing essays as an adolescent and received a high school award for creative writing and editorial work for the school magazine.

After finalizing his law diplomas, he graduated with an LL.M. in European Integration at Saarland University, Germany, in 1982, and with a Doctor of Law title from University of Geneva, Switzerland, in 1987.

He then took courses in psychology at the University of Geneva and interviewed a number of psychotherapists in Lausanne and Geneva, Switzerland. His interest was intensified through a hypnotherapy with an Ericksonian American hypnotherapist in Lausanne. This led him to the recovery and healing of his inner child.

After a second career as a corporate trainer and personal coach, Pierre retired in 2004 as a full-time writer, philosopher and consultant.

His nonfiction books emphasize a systemic, holistic, cross-cultural and interdisciplinary perspective, while his fiction works and short stories focus upon education, philosophy, perennial wisdom, and the poetic formulation of an integrative worldview.

Pierre is a German-French bilingual native speaker and writes English as his 4th language after German, Latin and French. He also reads source literature for his research works in Spanish, Italian, Portuguese, and Dutch. In addition, Pierre has notions of Thai, Khmer, Chinese, Japanese, and Vietnamese.

All of Pierre's books are hand-crafted and self-published, designed by the author. Pierre publishes via his Delaware company, Sirius-C Media Galaxy LLC, and under the imprints of IPUBLICA and SCM (Sirius-C Media).

The author's profits from this book are being donated to charity.

Contents

Introduction — 11
About Great Minds Series

Chapter One — 15
Short Biography

Chapter Two — 23
Fritjof Capra's Contributions to Holistic Science

Introduction	23
The Pioneer	29
The Systems Thinker	46
The Social Critic	65
The Pragmatist	69

Chapter Three — 81
The Tao of Physics

Review	82
Critique	93
Quotes	100

Chapter Four — 109
Green Politics

    Chapter 2 — 111

  Principles of a New Politics

    Chapter 9 — 114

  Possibilities for Green Politics in America: 1983

Chapter 10   117

Green Politics in the United States: 1986

Ten Key Values   117

1. Ecological Wisdom
2. Grassroots Democracy
3. Personal and Social Responsibility
4. Nonviolence
5. Decentralization
6. Community-based Economics
7. Postpatriarchal Values
8. Respect for Diversity
9. Global Responsibility
10. Future Focus

Chapter Five   123
Belonging to the Universe

3 Paradigms in Science and Theology (pp. 33-39)   125

Paradigms in science and society

New thinking and new values (pp. 73-77)   135

Chapter Six   141
The Turning Point
Review   142
Quotes   153

Chapter Seven   175
Uncommon Wisdom
Review   176
Quotes   193

CONTENTS

Chapter Eight   209
The Web of Life

| Review | 209 |
| Quotes | 222 |

Chapter Nine   249
The Hidden Connections

| Review | 250 |
| Quotes | 257 |

Chapter Ten   295
Steering Business Toward Sustainability

| Review | 296 |
| Quotes | 304 |

Chapter Eleven   309
The Science of Leonardo

| Review | 310 |
| Quotes | 318 |

Chapter Twelve   325
Learning from Leonardo

| Review | 326 |
| Quotes | 336 |

Chapter Thirteen   343
The Systems View of Life

| Review | 344 |
| Quotes | 347 |

Bibliography 361
German and French Editions of Fritjof Capra's Books

Personal Notes 363

Introduction
About Great Minds Series

We are currently transiting as a human race a time of great challenge and adventure that opens to us new pathways for rediscovering and integrating the perennial holistic wisdom of ancient civilizations into our modern science paradigm. These civilizations were thriving before patriarchy was putting nature upside-down.

Currently, with the advent of the networked global society, and systems theory as its scientific paradigm, we are looking into a different world, with a rise of 'horizontal' and 'sustainable' structures both in our business culture, and in science, and last not least on the important areas of psychology, medicine, and spirituality.

—A paradigm, from Greek 'paradeigma,' is a pattern of things, a configuration of ideas, a set of dominant beliefs, a certain way of looking at the world, a set of assumptions, a frame of reference or lens, and even an entire worldview.

While most of this new and yet old path has yet to be trotted, we cannot any longer overlook the changes that happen all around us virtually every day.

Invariably, as students, scientists, doctors, consultants, lawyers, business executives or government officials, we face problems today that are so complex, entangled and novel that they cannot possibly be solved on the basis of our old paradigm, and our old way of thinking. As Albert Einstein said, we cannot solve a problem on the same level of thought that created it in the first place— hence the need for changing our view of looking at things, the world, and our personal and collective predicaments.

What still about half a decade ago seemed unlikely is happening now all around us: we are rediscovering more and more fragments of an integrative and holistic wisdom that represents the cultural and scientific treasure of many ancient tribes and kingdoms that were based upon a perennial tradition which held that all in our universe is interconnected and interrelated, and that humans are set in the world to live in unison with the infinite wisdom inherent in creation as a major task for driving evolution forward!

It happens in science, since the advent of relativity theory, quantum physics and string theory, it happens in neuroscience and systems theory, it happens in molecular biology, and in ecology, and as a result, and because science is a major motor in society, it happens now with increasing speed in the industrial and the business world, and in the way people earn their lives and manifest their innate talents through their professional engagement.

And it happens also, and what this series of books is *inter alia* set to emphasize, in psychology and psychoanalysis. See, for example that Françoise Dolto (1908-1988), the famous French psychotherapist, while having been a member of the Freudian psychoanalytic school, has created an approach to healing psychotic children that was really unknown to the founder of psychoanalysis, Sigmund Freud.

More and more people begin to realize that we cannot honestly continue to destroy our globe by disregarding the natural law of self-regulation, both outwardly, by polluting air and water, and inside, by tolerating our emotions to be in a state of repression and turmoil.

Self-regulation is built into the life function and it can be found as a consistent pattern in the lifestyle of native peoples around the world. It is similar with our immense intuitive and imaginal faculties that were downplayed in centuries of darkness and fragmentation, and that now emerge anew as major key stones in a worldview that puts the *whole human* at the frontline, a human who uses their whole brain, and who knows to balance their emotions and natural passions so as to arrive at a state of inner peace and synergetic relationships with others that bring mutual benefit instead of one-sided egotistic satisfaction.

For lasting changes to happen, however, to paraphrase J. Krishnamurti, we need to change the thinker, we need to undergo a transformation that puts our higher self up as the caretaker of our lives, not our conditioned ego.

Hence the need to really look over the fence and get beyond social, cultural and racial conditioning for adopting an integrative and holistic worldview Th. at is focused on more than problem-solving.

What this book tries to convey is that taking the example of one of the greatest holistic thinkers of our time, we may see that it's not too late, be it for our planet or for us humans, our careers, our science, our collective spiritual advancement, and our scientific understanding of nature, and that we can thrive in a world that is surely more different in ten years from now than it was one hundred years in the past compared to now.

We are free to continue to feel like victims in this new reality, and wait for being taken care of by the state, or we may accept the state, and society, as human creations that will never be perfect, and venture into creating our lives and careers in accordance with our true mission, and based upon our real gifts and talents.

Let me say a last word about this series of books about great personalities of our time, which I came to call 'Great Minds' Collection. The books within this collection do not just feature books but *authors*, you may call them author reviews instead of book reviews, and they are more extensive also in highlighting the personal mission and autobiographical details which are to note for each author, including extensive quotes from their books.

Chapter One
Short Biography

Fritjof Capra is well-known and famed as one of the most important authors on new science and systems theory.

I found Capra's *Tao of Physics* in 1985, at a time when my life was in a complete reorientation. In this situation, Capra's books *The Tao of Physics* and *The Turning Point* reflected the turning point in my own life.

The impact of *The Tao of Physics* on my personal journey was comparable only to my discovery of the I Ching and Taoism, as well as the psychoanalytic teaching of the late *Françoise Dolto (1908-1988)*. Besides Capra's intellectual brilliance and exquisite use of language, it's the simplicity of his diction, and his unpretentious way to relate other people's achievements and remarkable traits with accuracy and tact that make Capra stand out not just as a scientist, but as an encyclopedic scholar. The fact that his books have become worldwide bestsellers over many years, and were translated in all major languages of the world shows his immense popularity and may also be a signal that his message is accepted by the intelligent strata of modern society.

The Tao of Physics asserts that both physics and metaphysics lead inexorably to the same knowledge, or are two visions of the universe that complement one another. The book was a door opener for many people while it was first regarded as a somewhat too daring perspective put in the world by a scientist. It is worthwhile to have a closer look at some biographical details which will help to better understand Fritjof Capra's stand in life, and his mission.

Fritjof Capra was born February 1st, 1939. His birth in the sun sign *Aquarius* may or not be considered an auspicious sign of his later career and his mission for ecology which reflects a foremost concern of the *Aquarius Age* into which we are currently heading. The son of the Austrian poet Ingeborg Capra-Teuffenbach, Capra graduated in 1966 at the University of Vienna with a doctorate in theoretical physics. He studied with Werner Heisenberg and researched and taught particle physics and systems theory at the University of Paris (1966–1968), the University of California, Santa Cruz (1968–1970), the Stanford Linear Accelerator Center (1970), Imperial College, London (1971–1974) and the Lawrence Berkeley Laboratory (1975–1988). While at Berkeley, he was a member of the Fundamental Fysiks Group, founded in May 1975 by Elizabeth Rauscher and George Weissmann, which met weekly to discuss philosophy and quantum physics. He has also taught at U.C. Santa Cruz, U.C. Berkeley, and San Francisco State University.

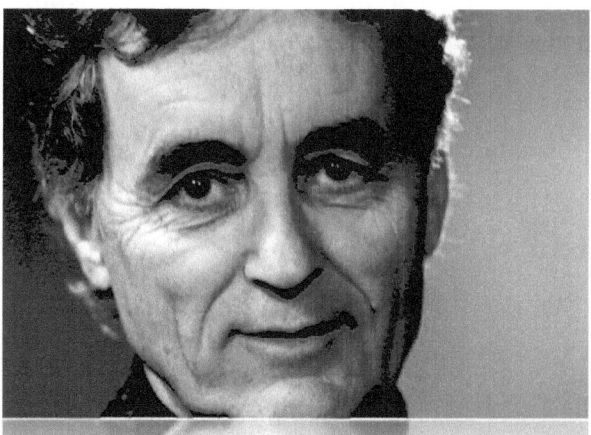

Capra is fluent in German, English, French and Italian, and both a Buddhist and a Catholic Christian. After touring Germany in the early 1980s, Capra co-wrote *Green Politics* with Charlene Spretnak in 1984.

He contributed to the screenplay for the movie *Mindwalk (1990)*, starring Liv Ullman, Sam Waterston and John Heard. The film is loosely based on his book, *The Turning Point (1987)*. The book was also the inspiration for a broad ad campaign called 'The Turning Point Project.' In the Fall of 2000, under the leadership of Jerry Mander and Andrew Kimbrell, this project produced full-page ads in USA Today and The New York Times criticizing nanotechnology.

Back in 1991, Capra co-authored *Belonging to the Universe* with David Steindl-Rast, a Benedictine monk who has been called a contemporary Thomas Merton. Steindl-Rast baptized Capra's daughter in a joint Christian-Buddhist ceremony. Using Thomas Khun's 'The Structure of Scientific Revolutions' as a stepping stone, their book explores the parallels between new paradigm thinking in science and religion that together offer what the authors consider to be remarkably compatible views of the universe.

Capra's mission more and more turned out to formulate a cultural critique of conventional linear thought and the mechanistic views of Descartes. Exposing Descartes' reductionist view that everything can be studied in parts for understanding the whole, he allows his readers to step back and look at the world through the eyes of systems research and complexity theory.

Capra is setting the grounds for change in many new theories, as for example in the systems view of life which is to be considered as the *theoretical framework of deep ecology*. This theory is only now fully emerging but it has its roots in several scientific fields that were developed during the first half of the twentieth century—organismic biology, gestalt psychology, ecology, general systems theory, and cybernetics.

Capra's ecological vision proposes modern society to abandon conventional linear thought and the mechanistic view of the universe, and develop a holistic science paradigm. He is a founding director of the *Center for Ecoliteracy* in Berkeley, California, which promotes ecology and systems thinking in primary and secondary education.

According to Capra, our economic and social problems such as unemployment, crime, pollution or global warming are the result of a perception crisis in modern society. A globally networked world cannot be understood any more within the framework of a reductionist and mechanistic science as it was practiced by Descartes and Newton, but needs to be transformed into a holistic and organic view of reality. Once this view is adopted, it will be obvious how many hidden connections there are between phenomena that the old worldview considers as separate, and how much in life, and in living systems, co-evolves by means of an often invisible interdependence.

As Capra explained in a lecture at Mill Valley School District, April 18, 1997, entitled *Creativity and Leadership in Learning Communities*, which was published by the Center for Ecoliteracy, ecoliteracy means to be 'ecologically literate' which means the understanding of ecological communities, also called ecosystems, and then using those principles for creating sustainable communities.

—Ecology is a term derived from the Greek word *oikos* (household); thus it conveys the study of the relationships between all members of the household called 'Earth.'

Ecological thinking is thus concerned with relationships, connectedness and context; in science it is called systems thinking.

In the same lecture he reports that one of the early insights of systems thinking was the realization that every living system is a network. At first, ecologists formulated the concepts of food chains and food cycles, and these were then expanded to the concept of the food web.

The *Web of Life*, a book that Capra published in 1997, is an old idea, which has been used by poets, philosophers, and mystics throughout the ages to convey their sense of the interwovenness and interdependence of all phenomena. In this sense, systems theory is actually a link back to the oldest of science traditions, which by itself proves that science, too, is *cyclic* as everything else in life.

In his leadership programs, Fritjof Capra stresses the fact that leaders could learn to understanding themselves using the systems view to being able to bring about emergence. This kind of leadership needs to be *team leadership,* not single leadership, as in all self-organizing systems leadership is distributed, and responsibility becomes 'a capacity of the whole.'

SHORT BIOGRAPHY

Leadership thus consists, according to Capra, in facilitating the emergence of new structures, and to incorporate the best of them into the organization's design.

—See also Peter Fritz Walter, Walter's Leadership Guide: Why Good Leadership Starts With Self-Leadership (2015) and The Leadership I Ching, 2nd Edition (2015).

There are other important facts about Capra that are perhaps lesser known, and partly explain why he has this phenomenal lucidity, while he works as a scientist and yet in his books by far surpasses the limitations of this profession and the much more limited worldview of most of his professional colleagues, except those on his own level of genius.

Capra wrote in his most autobiographical book, *Uncommon Wisdom (1989),* that he was raised in a quite matriarchal environment, an environment virtually deprived of males. He was raised by three women, and they were all single, for different reasons: his mother, his grandmother and his great grandmother. And they lived together with many animals on the big farm.

All this is important, I think, for understanding his basically *non-judgmental worldview* and his ability to understand people from ultra-orthodox to very liberal with the same generosity and magnanimity.

Capra is truly exceptional in this respect. This can be seen in *Uncommon Wisdom (1989)* which is a recollection of conversations with remarkable people, and at the same time a kaleidoscope of anecdotes that form the life of a truly lively and communicative human being.

The other noteworthy instance from Capra's life is his long involvement with the counterculture and his meeting with most of the celebrities of that culture, as for example *Timothy Leary, Terence McKenna, Gregory Bateson,* or *Ronald David Laing* and *Thomas Szasz*, the founders of the *antipsychiatry* movement.

Capra's extraordinary *human skills,* his ability to communicate across scientific disciplines together with an integrative mindset and attitude make him an important alternative figure in the mainstream science environment.

Capra is in my view one of the most important holistic thinkers of our times, and perhaps even *the* most important of our science philosophers today.

Chapter Two
Fritjof Capra's Contributions to Holistic Science

Introduction

Without holistic science having been developed to a point it is accepted by a wide range of scholars—while it is certainly not yet a mainstream paradigm—we could not assume at this point that energy science, a *science of the protoplasmatic energy field,* can emerge. This preparative work needed to be accomplished and it was Fritjof Capra who was one of the pioneers in this process that started in the 1960s and 70s, while in a wider context, we can say that it started in 1905, when Einstein wrote his first draft of special relativity theory.

I am going to point out in this chapter that besides Capra's intellectual brilliance and exquisite use of language, it's the simplicity of his literacy, and his *unpretentious* way to relate other people's achievements and remarkable traits with a certain modesty and apparently without jealousy—which is often to be found in the science establishment—, that make Capra stand out as a truly universal and encyclopedic scholar.

I see many connection points between Capra's systems research and my own more than twenty years of research on human emotions and sexual identity, the human energy field and quantum physics.

It was Capra who opened my eyes, back in the 1980s, to the importance of a *systems view of life*, and the need for all of us to contribute to formulating a new holistic science paradigm for the future. It was like a call for me, and if

Wilhelm Reich was one leg I was standing on in my career as a science writer, Capra was the other leg.

I also appreciate and always positively validate the rational and intelligent grasp that Capra had of Reich's scientific career and discoveries. Besides specialized researchers who are often two zealous to defend and try to rehabilitate the defamed Austrian medical doctor, I haven't found so far appreciative comments on Reich's scientific achievements in the publications of mainstream scientists.

Here too, Capra marks the exception to the rule, and I have greatly applauded his courage, as it's sometimes met with aggression, even today, when researchers show in which ways our science tradition constantly downgrades, betrays and defames its greatest avatars.

Others to be mentioned here are Paracelsus, Franz Anton Mesmer and Nikola Tesla; none of them has so far been really validated in mainstream science as giant pioneers of modern science, while in the rainbow press they are becoming increasingly popular, just like Reich.

One of the angular stones in Capra's scientific publications, and for which he has deserved a medal, is to have worked through the intricacies of the mechanistic, dualistic and reductionist science paradigm that reigned in our culture for the last four hundred years or so, and that came up as a deliberate opposition to the scientific dogmatism of the Christian Church. Capra showed the multiple convolutions and distortions that this science paradigm contains, and its twisted perception of the whole of the process of life. In fact, by contrasting the results this paradigm brings when comparing it with the systems view of life, the almost perverse manner in which it tears down to static elements the organic and systemic composure of life to a huge clockwork becomes evident even to the interested lay reader.

Capra invested a lot of work in this research as he retraced, in a detailed science-historical review, our present mainstream science paradigm back to its roots in ancient Greece, showing all the important bifurcations that led us to where we are today. Capra also lucidly shows and explains that it was *quantum physics* that triggered, quite brutally, the turndown of the Newtonian worldview, and the emergence of a temporary chaos situation in mainstream science that led, within the first two decades of the 20th century, to the early formulation of quantum theory. This new science paradigm may be solid today but we have to see how many years and decades of development it needed to be built, and by how many contributing scientists this was achieved.

The fundamental difference here is also one of *teamwork vs. single genius.* Capra showed that the special relativity theory was almost totally the work of one man, Albert Einstein, who formulated it back in 1905, and that that was about the last time when a single scientist could break open new land, while formerly in human scientific history, this was the rule.

Aristotle formulated a complete science methodology that later was taken over by the Church, and thus was valid for several hundreds of years without being questioned—and this despite the fact that it was *basically flawed*.

These times, as Capra showed, are definitely over, and today scientific progress is a matter of teamwork and repeated trial and error, and the gradual corroboration of theories through innumerable experiments that take place in different countries and cultures, thereby giving the variety of human intelligence its greatest chance to impact upon new and promising scientific developments.

Reading Capra, I found my intuition confirmed that not only our science tradition, but also education needs to be reformed. It is a gigantic task as we have to put together nothing short but a structural framework for the purpose of creating a *new reality*, a reality that will be *holistic and emotionally as well as erotically intelligent*.

In fact, it was at that time that I decided to devote myself to serving humanity unconditionally for helping to create this new reality and work for this mission diligently, based upon a transpersonal motivation.

Originally from Austria and brought up with German as his mother tongue, Capra learnt English so perfectly that from the moment he moved to Berkeley, California for his work as a quantum physicist, he wrote and published only in English. The parallels are obvious to Albert Einstein and Wilhelm Reich who equally were from Germanic origin and after their immigration to the United States only wrote and published in English.

And from their level of genius and originality, these three men can well be compared.

All these details are important in that they provide keys for understanding his ability to bridge over personal animosities and jealousy to get a variety of professionals and scientists into scientific dialogue with the same generosity and magnanimity.

A good example for Capra's communicative spirit is *Uncommon Wisdom (1989)*, a recollection of conversations with remarkable people, and at the same time a series of anecdotes told by somebody who is alive and communicative. Another noteworthy instance from Capra's life is his long standing love affair with the counterculture and his notable relations with most of the celebrities of that culture, as for example R.D. Laing, Timothy Leary, Terence McKenna, Gregory Bateson, and others.

Fritjof Capra

Uncommon Wisdom

Conversations with Remarkable People

New York, Bantam, 1989

The Pioneer

In this chapter, I shall trace Capra's exciting scientific journey from the mid 1980s to today, two decades of a steady course of action strongly focused upon bringing about a holistic and systemic science paradigm for today's industry needs and the future of human evolution.

In addition, Capra has become a pioneer in the field of what he came to call *ecological literacy*, and he is to be considered as one of the leading ecology experts in the world.

Besides, Fritjof Capra is a renowned consultant to government and industry, enjoying considerable reputation in the United States, Germany, Brazil, or Russia, and around the world; as such he is influential in the good sense, and for a good cause.

FRITJOF CAPRA AND THE SYSTEMS VIEW OF LIFE

Ralph Abraham writes in an online book review of one of Capra's latest books, *The Science of Leonardo (2007)*:

> While viewing an exhibition of his drawings in the mid 1990s, Capra decided to make a detailed study of his writings. As a scientist having acquired Italian language in childhood, he was able to study the recently transcribed and dated *Notebooks* of Leonardo, paying special attention to their scientific content. *The Science of Leonardo* is the outcome of that process. In his Introduction, Capra gives us a portrait of Leonardo as a systems thinker, the first modern scientist, pioneer of the experimental method—a century before Galileo and Bacon. In Part One, *Leonardo the Man*, Capra reviews Leonardo's life in Florence in the 1470s, in Milan from the 1480s, and in Rome from 1513 until his death. Capra outlines the key ideas of science from the Notebooks—a science of living forms, the movements of water, the forms of the living earth, macrocosm and microcosm, nature's machines, and the mystery of human life. In addition he presents Leonardo's highly original and little known contributions to mathematics—the geometry of proportions, the geometry of nature, the geometry of functions and curves, and the theory of continuous motions of curves, anticipating Leibniz. In the final two chapters of Part Two, *Leonardo the Scientist*, Capra details Leonardo's theory of knowledge, and his genius as a systems thinker. In the *Epilogue*, Capra sums up in six pages his view of Leonardo, and contrasts it with the various specialized scientific biographies published previously. This book is a thrilling read for fans of the history of science, and a must for contemporary systems thinkers.
>
> —http://www.ralph-abraham.org/reviews/capra-rvw.pdf

 An important discourse in the *Tao of Physics* was Capra's giving an important clue for the origins of our intellectual dualism, a subject that Joseph Campbell has widely covered in *Occidental Mythology (1991)*. Capra's scientific journey is directed against that dualism, and represents an attempt to overcome that schizoid split by showing that upon a deeper regard, a *synthesis* of left-brain or deductive Western thought and right-brain and inductive Eastern philosophy is the only intelligent way out of the dilemma caused by the stringent paradoxes of quantum mechanics.

What I came to call in my writings the *schizoid split* in the internal setup of Western culture, Capra calls it the division between spirit and matter:

> As the idea of a division between spirit and matter took hold, the philosophers turned their attention to the spiritual world, rather than the material, to the human / soul and the problems of ethics. These questions were to occupy Western thought for more than two thousand years after the culmination of Greek science and culture in the fifth and fourth centuries B.C.
>
> —Fritjof Capra, The Tao of Physics (1975/1984), 6-7.

Capra left no doubt that it was *Aristotle* who ultimately forged that dualism in our cultural credo for the next two thousand years:

> The scientific knowledge of antiquity was systematized and organized by Aristotle who created the scheme which was to

be the basis of the Western view of the universe for two thousand years. But Aristotle himself believed that questions concerning the human soul and the contemplation of God's perfection were much more valuable than investigations of the material world. The reason the Aristotelian model of the universe remained unchallenged for so long was precisely this lack of interest in the material world, and the strong hold of the Christian church which supported Aristotle's doctrines throughout the Middle Ages. (Id., 8)

And the next step, then, in the building of that cultural paranoia was the turn of events starting with the reductionist science philosophy of the French philosophers La Mettrie and René Descartes:

> The birth of modern science was preceded and accompanied by a development of philosophical thought which led to an extreme formulation of the spirit/matter dualism. This formulation appeared in the seventeenth century in the philosophy of René Descartes who based his view of nature on a fundamental division into two separate and independent realms: that of mind (res cogitans), and that of matter (res extensa). The Cartesian division allowed scientists to treat matter as dead and completely separate from themselves, and to see the material world as a multitude of different objects assembled into a huge machine. (Id.)

Capra showed the missing link between our separative and individualistic worldview and its historical origins; it explains why we are torn up inside, fragmented and unwhole (unholy):

This inner fragmentation mirrors our view of the world outside, which is seen as a multitude of separate objects and events. The natural environment is treated as if it consisted of separate parts to be exploited by different interest groups. The fragmented view is further extended to society, which is split into different nations, races, religions and political groups. (Id.)

The danger of fragmentation, Capra explains, is that we try to find absolute points of reference behind each of our fragmented concepts, and we do this probably unconsciously in an attempt to heal our inner split. Yet ultimately by doing so we bring about a *distorted perception* by taking the proverbial finger that points to the moon, for the moon.

In addition, facing the paradoxical behavior of electrons in the quantum world, Capra asked the intelligent question why Westerners are confused, and even shocked when encountering a paradox or simply a completely illogical behavior, while in Eastern philosophies and religions, paradoxes are a recurring feature? Ancient Indian philosophy, for example, is very comfortable with paradoxes, so are the Chinese and Japanese philosophical traditions.

The Zen tradition, derived from its original Chinese root philosophy (where it was called *Chan Buddhism*), is very fond of putting the stress on the paradox for a simple reason: the paradox teaches us the limitations of rational thinking and thereby shows us the relativity of a purely rational worldview. By seeing our obvious limitations, we can go beyond them and develop a more holistic, integrative worldview, a worldview namely that gives the necessary space for the *irrational*, the *fantastic*, the *imaginative* and the *scurrilous* in nature, and also in our human nature. Without the latter, humor, for example, as an expression of true humanity, is not possible.

Capra makes it very clear that we cannot stay with the old Newtonian demons:

> The mechanistic view of nature ... is closely related to a rigorous determinism. The giant cosmic machine was seen as being completely causal and determinate. All that happened had a definite cause and gave rise to a definite effect, and the future of any part of the system could—in principle—be predicted with absolute certainty if its state at any time was known in all details. (...) The philosophical basis of this rigorous determinism was the fundamental division between the I and the world introduced by Descartes. As a consequence of this division, it was believed that the world could be described objectively, i.e., without ever mentioning the human observer, and such an objective description of nature became the ideal of all science. (Id., 45)

The result is that we discarded nature out of science and by doing this, created a *fundamentally nature-hostile science*, a science that is about to destroy us by destroying our planet. This science reflected the sexist cultural bias. The cultural norm of *male supremacy* led to a course of violence that slowly but definitely suffocates us. Capra writes:

Western society has traditionally favored the male side rather than the female. Instead of recognizing that the personality of each man and of each woman is the result of an interplay between male and female elements, it has established a static order where all men are supposed to be masculine and all women feminine, and it has given men the leading roles and most of society's privileges. This attitude has resulted in an over-emphasis of all the yang—or male—aspects of human nature: activity, rational thinking, competition, aggressiveness, and so on. The yin—or female—modes of consciousness, which can be described by words like intuitive, religious, mystical, occult, or psychic, have constantly been suppressed in our male-oriented society. (Id., 133)

This *biased perception of reality*, which is jeopardizing the harmony between the male and the female principle can be seen throughout Western philosophy, in its abysmal dualism, which lacks the fundamental ability to find the synthesis that Oriental thought is so apt to establish.

Capra agrees with the Eastern view that says all opposites are complementary and 'merely different aspects of the same phenomenon.' Capra wistfully remarks that in the East, 'a virtuous person is therefore not one who undertakes the impossible task of striving for the good and eliminating the bad, but rather one who is able to maintain a dynamic balance between good and bad.'

When you look at the *Tao of Physics* from this perspective, from the big picture behind the many fuzzy details of quantum physics, you will see that Capra's deeper message in this revolutionary book goes way beyond a redefinition of modern physics. Capra has prepared the ground in this earliest of his books for the giants to come, and he has given birth to the first giant, *The Turning Point (1982/1987)*, about a decade after publishing *The Tao of Physics (1975)*.

While *The Tao* remains Capra's most popular book it is perhaps not his best book. The trick was that he developed the original idea further and found something like a new holistic concept for all sciences, but never arrogated himself to label it fashionably as 'A Theory of Everything.' Capra calls his new concept 'ecological,' and while he has not invented that term, he surely has given a much broader content to it than it had ever before.

The Turning Point is one of Capra's most important books, and truly it was a turning point also in Capra's own life. In this book, he extrapolated the holistic concepts developed in the *Tao* to the whole of international culture.

Only a thinker who is logically precise, knowledgeable about science history, and who has a *metarational and integrated* perception of the universe could do such a giant work. As a matter of coincidence, this book was marking also a turning point in my own life, and I found it, much like a blessing, during a time of virulent contradictions and emotional turmoil in my life, back in 1985.

The following quote shows the general direction that Capra took after the publishing of this book, and that will be especially present in his two subsequent books, *The Web of Life (1996/1997)*, and *Hidden Connections (2002)*.

It has been called the *systems view*; it simply is a sound holistic science paradigm that can be practically applied to all scientific research, and that promises to bring about scientific, social and later political results that are in accordance with human dignity, fostering the expansion of human consciousness and evolution, while respecting sustainability and other ecological realities.

The solutions namely will be different from those we had in the past because they will be integrated and sustainable, and this both in the fields of science and culture:

> These problems (...) are systemic problems, which means that they are closely interconnected and interdependent. They cannot be understood within the fragmented methodology characteristic to our academic disciplines and government agencies. Such an approach will never solve any of our difficulties but will merely shift them around in the complex web of social and ecological relations. A resolution can be found only if the structure of the web is changed, and this will involve profound transformations of our social institutions, values, and ideas.
>
> —Fritjof Capra, The Turning Point (1982/1987), 6

Capra stresses that the systems view is not 'just theory,' but bears a direct significance for our daily life, and our daily problems. Contrary to many other scientists from the so-called 'exact' scientific disciplines, his thought is extraordinarily synthetic which makes him sense shifts and developments in society long before they actually happen.

Then, following his intuition, he puts his sharp rational mind in the forefront so as to collect and arrange the information he needs to elucidate, with the end goal of deploying what he intuitively anticipates. This is fully in accordance with Einstein's famous saying that a problem can never be solved on the level on which it was created. In fact, it's only through creative thinking and intuition that we can find new solutions to our old problems, because we then situate the thinker on a different level of perspective.

This can be seen in the ingenious manner Capra puts spotlights on trends and philosophical movements of long ago, to show the potential they have for shifting our view and preparing new ground for alternative, effective and unconventional solutions that will be the mainstream solutions in the future.

Heraclites is one of the earliest geniuses who showed us the way to go, but he was not followed. Instead Western science was to slavishly follow Aristotle, and in the East, the same happened when Lao-tzu was shunned by Chinese thinkers, giving the preference to the pedant, moralist and hair-splitting Confucius.

And not coincidentally so, presently, while we got all kinds of fancy developments in modern science in the aftermath of *What the Bleep Do We Know!?* and *Theories of Everything* virtually sprouting out of every university faculty, we are rather in a restorative and ultra-conservative trend that does not really welcome a systemic and ecology-based science paradigm.

I can't think of any moment in history when the flourishing of science and the arts, material abundance as well as prosperity, and the intellectual prowess of the great nations was in *greater contradiction* with the setup of its reigning political power structures. One important area where the reigning paradigm is presently shifting is *psychology*. This is only *now* really apparent, where we can count the books written about what today is called *Energy Psychology*, but at the time Capra authored *The Turning Point*, this was unthinkable. I have seen it myself when, after finalizing my *doctor of law* degree at the University of Geneva, in 1987, I was shifting majors and started to study psychology. And the first semester consisted of sixty percent statistics, and a few lectures on basic psychological terminology and ways of research. There was no word about psychoanalysis, no word about any holistic or systemic area of research, and it was boring me to death. I always thought that law is a dry subject to study but after having peeked in psychology, I can say that law is one of the most colorful and passionate subjects I have ever studied!

Capra explains why the systems view of life will have a profound impact upon psychology:

> As in the new systems biology, the focus of psychology is now shifting from psychological structures to the underlying

processes. The human psyche is seen as a dynamic system involving a variety of functions that systems theorists associate with the *phenomenon of self-organization*. Following Jung and Reich, many psychologists and psychotherapists have come to think of mental dynamics in terms of a flow of energy, and they also believe that these dynamics reflect an intrinsic intelligence—the equivalent of the systems concept of mentation—that enables the psyche not only to create mental illness but also to heal itself. Moreover, inner growth and self-actualization are seen as essential to the dynamics of the human psyche, in full agreement with the emphasis on self-transcendence in the systems view of life. (Id., 407)

In fact, one of Capra's friends is Stanislav Grof, and with Grof he discussed many of the topics of psychology/psychiatry he writes about. I got this information not only from the huge footnote section in the *Turning Point*, but also from his insightful book *Uncommon Wisdom*, in which he published interviews with leading edge personalities from all walks of life, and that stands as an example for Capra's extraordinary communication abilities. After having read *Getting Well Again (1978)*, I can say that Capra did not promise too much with his summary of the revolutionary work of these doctors. All the cutting-edge information about the book and the approach is concisely presented by Capra:

> The popular image of cancer has been conditioned by the fragmented world view of our culture, the reductionist approach of our science, and technology-oriented practice of medicine. Cancer is seen as a strong and powerful invader / that strikes the body from outside. There seems to be no

hope of controlling it, and for most people cancer is synonymous with death. Medical treatment—whether radiation, chemotherapy, surgery, or a combination of these—is drastic, negative, and further injures the body. Physicians are increasingly coming to see cancer as a systemic disorder; a disease that has a localized appearance but has the ability to spread, and that really involves the entire body, the original tumor being merely the tip of the iceberg. (Id., 388-389)

What many physicians try to veil is the fact that the strangeness of the current cancer cure has nothing specific about it, and can be well explained, and criticized, by looking through its mechanistic nature. It is a mechanistic and inhuman approach, not directed at true healing, but *medical business*, a worldwide money-making machine of which all the huge profits go in the hand of a few multinationals that use a legion of uncritical doctors as their brave business consultants.

Capra writes in slightly more hopeful terms, when he reports the Simonton approach to cancer healing. But the fact alone that the Simontons are successful in their approach shows with the best possible evidence that they must be right somehow:

> One of the main aims of the Simonton approach is to reverse the popular image of cancer, which does not correspond to the findings of current research. Modern cellular biology has shown that cancer cells are not strong and powerful but, on the contrary, weak and confused. They do not invade, attack, or destroy, but simply overproduce. A cancer begins with a cell that contains incorrect genetic information because it has been damaged by harmful substances or other environmental influences, or simply because the organism will occasionally produce an imperfect cell. The faulty information will prevent the cell from functioning normally, and if this cell reproduces others with the same incorrect genetic makeup, the result will be a tumor composed of a mass of these imperfect cells. Whereas normal cells communicate effectively with their environment to determine their optimal size and rate of reproduction, the communication and self-organization of malignant cells are impaired. As a result they grow larger than healthy cells and reproduce recklessly. Moreover, the normal cohesion between cells may weaken and malignant cells may / break loose from the original mass and travel to other parts of the body to form new tumors—which is known as metastasis. In a healthy organism the immune system will recognize abnormal cells and destroy them, or at least wall them off so they cannot spread. But if for some reason the immune system is not strong enough, the mass of faulty cells will

continue to grow. Cancer, then, is not an attack from without but a breakdown within. (Id., 389-390)

What I may add here to Capra's analysis is that this 'popular image of cancer' is not the result of folk wisdom, or folk delusion, but rather of *folk hypnosis*. The general public knows intuitively very well that what the official rhetoric says about cancer is not true, but what can and do they do against the medical establishment? What it does is to keep people sick because it's on the back of sick people, and not on the back of healthy, and critical people that it makes its return of investment. In fact, the public is brainwashed by a medical propaganda that has no parallel in human history and that has put the image of cancer as a 'killer disease' in the minds of all and everybody. It's not human feeling and natural intuition of the 'common man' that has created this standard metaphor of the hopeless and passive patient who is 'innocently executed' by a terminal disease.

It's a myth through and through, but it can spread like a virus because of the apathy of most consumer-citizens to see through the veil of lies they are presented every day in the media; that is the price they pay for their eternal passivity to find out for themselves where the truth is.

> The Simontons and other researchers have developed a psychosomatic model of cancer that shows how psychological and physical states work together in the onset of the disease. Although many details of this process still need to be clarified, it has become clear that the emotional stress has two principal effects. It suppresses the body's immune system and, at the same time, leads to hormonal imbalances that result in an increased production of abnormal cells. Thus optimal conditions for cancer growth are created. The production of malignant cells is enhanced precisely at a time when the body is least capable of destroying them. As far as the personality configuration is concerned, the individual's emotional states seem to be the crucial element in the development of cancer. The connection between cancer and emotions has been observed for hundreds of years, and today there is substantial evidence for the significance of specific emotional states. These are the result of a particular life history that seems to be characteristic of cancer patients. Psychological profiles of such patients have been established by a number of researchers, some of whom were even able to predict the incidence of cancer with remarkable accuracy on the basis of these profiles. (Id., 391)

The Systems Thinker

Here I shall discuss the later Capra, the pioneer in systems research, and I personally believe his best books are his later ones, *The Web of Life (1996/1997)* and *The Hidden Connections (2002)*.

Why, one may want to ask? Because it was in these books, and not in his earlier productions that Capra really defined his approach to ecology, thereby making ecology, or *deep ecology*, a concept that is part of a *new science paradigm*, powerfully introduced and promoted by one of the most important science theorists of our times.

What is deep ecology and why do we need it? Capra writes in *The Web of Life*:

> Whereas the old paradigm is based on anthropocentric (human-centered) values, deep ecology is grounded in ecocentric (earth-centered) values. It is a worldview that acknowledges the *inherent value of nonhuman life.*
>
> Such a deep ecological ethics is urgently needed today, and especially in science, since most of what scientists do is not life-furthering and life-preserving but life-destroying.
>
> With physicists designing weapon systems that threaten to wipe out life on the planet, with chemists contaminating the global environment, with biologists releasing new and unknown types of microorganisms without knowing the consequences, with psychologists and other scientists torturing animals in the name of scientific progress—with all these activities going on, it seems most urgent to introduce 'ecoethical' standards into science. (11)

This book's quest is enormous, in that it requires modern science to fundamentally shift its regard upon nature, and upon living. Our regard upon nature has been conditioned by patriarchy since about five thousand years, and it's a rather defensive, distorted, if not completely schizoid regard, so much the more as both our mainstream religious paradigm and the Cartesian shift of science in the 17th and 18th centuries have contributed to a *deeply reductionist view of nature.* Capra looked back in history and found amazing early intuitions and truths propagated by our great thinkers, poets and philosophers, such as for example Immanuel Kant, Goethe or William Blake. He writes:

> The understanding of organic form also played an important role in the philosophy of Immanuel Kant, who is often considered the greatest of the modern philosophers. An idealist, Kant separated the phenomenal world from a world of 'things-in-themselves.' He believed that science could offer only mechanical explanations, but he affirmed that in

areas where such explanations were inadequate, scientific knowledge needed to be supplemented by considering nature as being purposeful. (Id., 21)

On the same line of thinking, Capra investigated what the earth, the globe, the planet means for us today, and why our science and technologies are hostile to it and little caring for its preservation?

He found conclusive answers in ancient traditions that fostered what we call today a Gaia worldview, a respectful attitude toward the earth, the mother, the *yin* energy and generally values associated with the female side of living:

> The view of the Earth as being alive, of course, has a long tradition. Mythical images of the Earth Mother are among the oldest in human religious history. Gaia, the Earth Goddess, was revered as the supreme deity in early, pre-Hellenic Greece. Earlier still, from the Neolithic through the Bronze Ages, the societies of 'Old Europe' worshiped numerous female deities as incarnations of Mother Earth. (Id., 22)

This is how Capra, always grounded in common sense and meaningful retrospection introduces the novice reader to the concept of systems research or the *systems view of life*.

Historically we can observe a certain evolution in post-matriarchal thought, which was naturally systemic, from the *Atomistic Worldview* (Democritus), over the *Cartesian Worldview* (Newton, La Mettrie, René Descartes) and *Relativistic Worldview* (Einstein, Planck, Heisenberg), to the *Systemic Worldview* (Bohm, Bateson, Grof, Capra,

Laszlo, etc.) and the *Holographic Worldview* (Talbot, Goswami, McTaggart, etc.).

In all systems, we have to deal with different levels of complexity that are woven in each other, thus rendering it almost impossible to dissect parts of the system for closer research without distorting the whole of our research results. This means that, contrary to earlier vivisectionist science, we have to leave the system intact and focus our research onto the *whole* of it—which makes research very complex by definition.

Hence, we had to develop a new mathematics, which is called the *mathematics of complexity*, for dealing with the high complexity levels in living systems. This also means that mere analysis is more or less dysfunctional for inquiring into the natural logic of living systems. Capra explains:

> According to the systems view, the essential properties of an organism, or living system, are properties of the whole, which none of the parts have. They arise from the interactions and relationships among the parts. These properties are destroyed when the system is dissected, either physically or theoretically, into isolated elements. Although we can discern individual parts in any system, these parts are not isolated, and the nature of the whole is always different from the mere sum of its parts. (...) The great shock of twentieth-century science has been that systems cannot be understood by analysis. The properties of the parts are not intrinsic properties but can be understood only within the context of the larger whole. Thus the relationship between the parts and the whole has been reversed. (Id., 29)

Capra defines the *Web of Life* as 'networks within networks.'

> At each scale, under closer scrutiny, the nodes of the network reveal themselves as smaller networks. We tend to arrange these systems, all nesting within larger systems, in a hierarchical scheme by placing the larger systems above the smaller ones in pyramid fashion. But this is a human projection. In nature there is no 'above' or 'below,' and there are no hierarchies. There are only networks nesting within other networks. (Id., 35)

In fact, living systems are not, as our governmental and societal organization, hierarchical, but network-based, and thus expanding not up-to-down but *horizontally* by 'neuronally' linking segments to larger molecular structures that distribute information instantaneously over the whole of the network. You can also say that a living network is a system of *total information sharing* where there is not one single molecule that is uninformed at any point in time and space.

The fact that horizontal networks are *nested within other horizontal networks,* while the different networks all possess different levels of complexity makes research unendingly complex.

This is why high-performance computers have greatly aided to developing systems theory. But the most revolutionary insight here is that our usual habit of dissecting parts of a whole for further scrutiny and scientific investigation does not work with living systems. Why is this so?

Ultimately—as quantum physics showed so dramatically—there are no parts at all. What we call a part is merely a pattern in an inseparable web of relationships. Therefore the shift from the parts to the whole can also be seen as a shift from objects to relationships. (Id., 37)

Hence, our *approach to scientific investigation* must shift from an object-based to a relationship-based research approach when we deal with living systems.

This requires the researcher to change their inner setup; this is exactly what quantum physics revealed to us, that is, the observer's belief system will be reflected in the outcome of the research, as it's part of reality, and not to be dissected from it.

And there is one more crucial element in systems research that Capra explains and elucidates, the fact that in approaching quantum reality, and organic behavior, we have to learn the mathematics of probability. What is probability? It is *approximation of behavior*.

Dealing with approximations means that we leave behind the certainty principle and venture into what Heisenberg called the *uncertainty principle.*

Giving up certainty triggers fear. This fear was vividly described by Max Planck and Heisenberg when the paradigm began to shift and quantum physics slowly but definitely began to undermine Euclidian geometry and Newtonian assuredness. Why has our certainty about the universe been undermined? Well, when we look at Hindu philosophy and ancient Chinese science, certainty was actually never an element of holistic perennial science, but only a part of fragmented modern science.

Once we abandon certainty, we begin to grasp the notion of *approximation*, and of probability, and accordingly we will shift our mathematical constructs.

> What makes it possible to turn the systems approach into a science is the discovery that there is approximate knowledge. This insight is crucial to all of modern science. The old paradigm is based on the Cartesian belief in the certainty of scientific knowledge. In the new paradigm it is recognized that all scientific concepts and theories are limited and approximate. Science can never provide any complete and definite understanding. (Id., 41)

The next important centerpoint in the *Web of Life* is the introduction of the notion of *open systems*. What is an open system? Capra explains:

> Unlike closed systems, which settle into a state of thermal equilibrium, open systems maintain themselves far from

equilibrium in this 'steady state' characterized by continual flow and change. (Id., 48)

Living systems are *open systems*, and not closed systems which means that their main characteristic is *change and flow*, not continuity and static behavior. They are *far from equilibrium*, which is the single most revolutionary discovery of systems research.

This means living systems are constantly struggling against decay. And decay here means equilibrium. This is a very important insight as when we extrapolate it from living systems to metaphysical reality, we see that it applies also to human beings, and even to religions. When we are settled, we are dead; this is what it all boils down to. And this insight from systems research may help us to survive in a state far from equilibrium, putting our assuredness or fake assuredness away, to stay with probability, the *beginner's mind,* as it's wistfully expressed in Zen.

I have stressed in all my publications the importance of understanding the nature of the universe as a *patterned universe*, showing the importance in nature of *patterned intelligence,* or *patterned organization*. What are patterns? Capra explains the importance of pattern when he explores the meaning of *self-organization,* which is one major characteristic of living systems:

> To understand the phenomenon of self-organization, we first need to understand the importance of pattern. The idea of a pattern of organization—a configuration of relationships

characteristic of a particular system—became the explicit focus of systems thinking in cybernetics and has been a crucial concept ever since. From the systems point of view, the understanding of life begins with the understanding of pattern. (Id., 80)

For describing the nature of *patterned systems* we need to change or for the least upgrade our basic toolset of scientific investigation. Capra explains:

> In the study of structure we measure and weigh things. Patterns, however, cannot be measured or weighed; they must be mapped. To understand a pattern we must map a configuration of relationships. In other words, structure involves quantities, while pattern involves qualities. (Id., 81)

This requires a radical change in our scientific thinking because traditionally Cartesian science was quantity-based and measure-oriented, while *systemic science is quality-based and relationship-oriented,* which Capra exemplifies when looking at the properties involved in the scientific focus of both static and systemic science theory:

> Systemic properties are properties of pattern. What is destroyed when a living organism is dissected is its pattern. The components are still there, but the configuration of relationships among them—the pattern—is destroyed, and thus the organism dies. (Id.)

An important self-regulatory function in living systems are *feedback loops*. Without feedback loops, living systems could not be self-organizing. Capra explains:

Because networks of communication may generate feedback loops, they may acquire the ability to regulate themselves. For example, a community that maintains an active network of communication will learn from its mistakes, because the consequences of a mistake will spread through the network and return to the / source along feedback loops. Thus the community can correct its mistakes, regulate itself, and organize itself. Indeed, self-organization has emerged as perhaps the central concept in the systems view of life, and like the concepts of feedback and self-regulation, it is linked closely to networks. The pattern of life, we might say, is a network pattern capable of self-organization. This is a simple definition, yet it is based on recent discoveries at the very forefront of science. (Id., 82-83)

Another requirement in understanding living systems is to focus upon the intrinsic quality of living systems as *nonlinear systems* that require, to be understood, an equally *nonlinear* mathematical approach. One early realization of mathematical nonlinearity was the introduction of the fractal in mathematics. In fact, in my exchanges with the Swiss mathematician Peter Meyer who was the collaborator of Terence McKenna for the realization of the *Timewave Zero* calculus as a part of *Novelty Theory*, I learnt that time is a fractal. Capra explains:

> The great fascination exerted by chaos theory and fractal geometry on people in all disciplines—from scientists to managers to artists—may indeed be a hopeful sign that the isolation of mathematics is ending. Today the new mathematics of complexity is / making more and more people realize that mathematics is much more than dry

formulas; that the understanding of pattern is crucial to understand the living world around us; and that all questions of pattern, order, and complexity are essentially mathematical. (Id., 152, 153)

After having elucidated that systems research involves a *process-based scientific approach* rather than an object-based one, Capra presents the perhaps most important research topic in this book, the *reinvestigation of cognition* based on the insights from systems research. Capra pursues:

> The identification of mind, or cognition, with the process of life is a radically new idea in science, but it is also one of the deepest and most archaic intuitions of humanity. In ancient times the rational human mind was seen as merely one aspect of the immaterial soul, or spirit. (Id., 264)

In fact, the whole debate about information processing, vividly criticized in the early writings of think tank Edward de Bono, and the even larger debate about cybernetics make it all clear that *cognition* is currently in a process of profound reevaluation:

> The computer model of cognition was finally subjected to serious questioning in the 1970's when the concept of self-organization emerged. (...) These observations suggested a shift of focus—from symbols to connectivity, from local rules to global coherence, from information processing to the emergent properties of neural networks. (Id., 266)

In my scientific exploration of emotions, I have revisited our scientific grasp of emotions as it was *cognized in a fragmented and reductionist manner* under the Cartesian science paradigm.

—See Peter Fritz Walter, Integrate Your Emotions (2014).

Capra comprehensively explains that emotions are not singular elements but coherently organized within a patterned system in which cognition and response are intertwined in a self-regulatory and organic whole:

> The range of interactions a living system can have with its environment defines its 'cognitive domain.' Emotions are an integral part of this domain. For example, when we respond to an insult by getting angry, that entire pattern of physiological processes—a red face, faster breathing, trembling, and so on—is part of cognition. In fact, recent research strongly indicates that there is an emotional coloring to every cognitive act. (Id., 269)

The most important fact that systems theory teaches us about cognition is that it does not at all work like a computer processes information.

Information processing, already years ago in the words of Edward de Bono has been called a preoccupation of Western scientists, and this obsession was not justified because our brain does not process information as a computer does. Capra explains why:

> A computer processes information, which means that it manipulates symbols based on certain rules. The symbols are distinct elements fed into the computer from outside, and during the information processing there is no change in the structure of the machine. The physical structure of the

> computer is fixed, determined by its design and construction. The nervous system of a living organism ... interacts with its environment by continually modulating its structure, so that at any moment its physical / structure is a record of previous structural changes. The nervous system does not process information from the outside world but, on the contrary, brings forth a world in the process of cognition. (Id., 274-275)

Capra then answers to the debate about artificial intelligence and the myths it creates in the minds of masses of people:

> A lot of confusion is caused by the fact that computer scientists use words such as intelligence, memory, and language to describe computers, thus implying that these expressions refer to the human phenomena we know well from experience. This is a serious misunderstanding. For example, the very essence of intelligence is to act appropriately when a problem is not clearly defined and solutions are not evident. Intelligent human behavior in such situations is based on common sense, accumulated from lived experience. Common sense, however, is not available to computers because of their blindness of abstraction and the intrinsic limitations / of formal operations, and therefore it is impossible to program computers to be intelligent. (Id., 275-276)

True intelligence is *contextual*, as language is. No computer can understand meaning. A rat's intelligence is a million times closer to that of man than that of the most powerful and sophisticated computer. Capra notes:

> The reason is that language is embedded in a web of social and cultural conventions that provides an unspoken context of meaning. We understand this context because it is common sense to us, but a computer cannot be programmed with common sense and therefore does not understand language. (Id., 276)

Capra is known to be one of the finest ecologists on the globe, and he often travels for giving ecological advice to the governments and non-governmental agencies. He has put a stress on *sustainability*, a term that was introduced in the early 1980s by Lester Brown, founder of the *Worldwatch Institute*. He defined a sustainable society as one that is able to satisfy its needs without diminishing the chances of future generations.

Thus, a system is sustainable when it's not only functional but also well integrated in a *greater continuum* so that it has a good prognosis for survival, for continuity. Capra writes:

> Partnership is an essential characteristic of sustainable communities. The cyclical exchanges of energy and resources in an ecosystem are sustained by pervasive cooperation. Indeed, we have seen that since the creation of the first nucleated cells over two billion years ago, life on Earth has proceeded through ever more intricate arrangements of cooperation and coevolution. Partnership—the tendency to associate, establish links, live inside one another, and cooperate—is one of the hallmarks of life. (Id., 278)

Partnership and cooperation were indeed alien words under patriarchy but they were imbedded in the pre-patriarchal civilizations, such as the *Minoan Civilization*, and thus what we are in today is a return to the sources.

Unfortunately most of our governments have a cynical attitude when it goes to recognize the need to protect our earth from being destroyed by ruthless and non-ecological technologies. Capra is quite outspoken here:

> The 1991 war in the Persian Gulf, for / example, which killed hundreds of thousands, impoverished millions, and caused unprecedented environmental disasters, had its roots to a large extent in the misguided energy policies of the Reagan and Bush administrations. (Id., 299-300)

And hopefully so, the fact that dedicated ecologists such as Dr. Capra are today traveling the world to consult governments for drafting more ecologically friendly policies will contribute to change both our administrative and business thinking; it will hopefully prepare the ground of sound ecology, partnership, cooperation and respectful communication that crosses national and cultural borders. This is what Capra says about this subject in *The Hidden Connections (2002)*:

> Organizations need to undergo fundamental changes, both in order to adapt to the new business environment and to become ecologically sustainable. This double challenge is urgent and real, and the recent extensive discussions of organizational change are fully justified. However, despite

these discussions and some anecdotal evidence of successful attempts to transform organizations, the overall track record is very poor. (Id., 99)

The Hidden Connections is perhaps the *culmination point* of Capra's systemic journey and his formulation of a viable holistic science paradigm.

As this chapter is not only about Capra's research but also about the man, the human Fritjof Capra, I would like to quote as much of personal information about him as I found in his books. To begin with, at the very onset of *The Hidden Connections*, Capra reveals an important information about himself and his unusual development as a scientist:

> My extension of the systems approach to the social domain explicitly includes the material world. This is unusual, because traditionally social scientists have not been very interested in the world of matter. Our academic disciplines have been organized in such a way that the natural sciences deal with material structures while the social sciences deal with social structures, which are understood to be, essentially, rules of behavior. In the future, this strict division will no longer be possible, because the key challenge of this new century—for social scientists, natural scientists and everybody else—will be to build ecologically sustainable communities, designed in such a way that their technologies and social institutions—their material and social structures—do not interfere with nature's inherent ability to sustain life. (xix)

Capra starts, systemically sound, with the cell, noting that the simplest living system is the cell, and especially, the *bacterial cell*. Then Capra looks at what *membranes* are, and what they do, and this is highly revealing, and teaches an important lesson about relationships. I haven't found this insightful metaphor anywhere else, and it showed me right at the start of the book that it's going to be highly substantial lecture:

> A membrane is very different from a cell wall. Whereas cell walls are rigid structures, membranes are always active, opening and closing continually, keeping certain substances out and letting others in. The cell's metabolic reactions involve a variety of ions, and the membrane, by being semipermeable, controls their proportions and keeps them in balance. Another critical activity of the membrane is to continually pump out excessive calcium waste, so that the calcium remaining within the cell is kept at the precise, very low level required for its metabolic functions. All these activities help to maintain the cell as a distinct entity and protect it from harmful environmental influences. Indeed, the first thing a bacterium does when it is attacked by another organism is to make membranes. (Id., 8)

The next important point to understand how nature 'thinks' is the cell's *metabolism*, the network that serves recycling. Capra succinctly elaborates:

> When we take a closer look at the processes of metabolism, we notice that they form a chemical network. This is another fundamental feature of life. As ecosystems are understood in terms of food webs (networks of organisms), so organisms are viewed as networks of cells, organs and organ systems,

and cells as networks of molecules. One of the key insights of the systems approach has been the realization that the network is a pattern that is common to all life. Wherever we see life, we see networks. (…) The metabolic network of a cell involves very special dynamics that differ strikingly from the cell's nonliving environment. Taking in nutrients from the outside world, the cell sustains itself by means of a network of chemical reactions that take place inside the boundary and produce all of the cell's components, including those of the boundary itself. (Id., 9)

I shall leave out the long passages in which Capra explains essential contributions of systems researchers such as Varela, Maturana or Prigogine, as this would render this chapter unreadable, and restrict myself to a few remarks in which I try to describe the core of systems research that Capra unfolds in this book:

The starting point for this is the observation that all cellular structures exist far from equilibrium state—in other words, the cell would die—if the cellular metabolism did not use a continual flow of energy to restore structures as fast as they are decaying. This means that we need to describe the cell as an open system. Living systems are organizationally closed—they are autopoietic networks—but materially and energetically open. (Id., 13)

What is chaos? What is order? We all have some preconceptions here. Chaos is not chaos, but ordered chaos, and thus not just random. Here, Capra explains in more detail what self-organization actually does:

Th[e] spontaneous emergence of order at critical points of instability is one of the most important concepts of the new understanding of life. It is technically known as self-organization and is often referred to simply as emergence. It has been recognized as the dynamic origin of development, learning and evolution. In other words, creativity—the generation of new forms—is a key property of all living systems. And since emergence is an integral part of the dynamics of open systems, we reach the important conclusion that open systems develop and evolve. Life constantly reaches out into novelty. (Id., 14)

The next great error most of us are caught in and that is the result of left-brain thinking is the distinction we have taken between humans and animals when it is about cognition. The truth is that we are not much more intelligent than Gorillas, only a little more, to be precise just a factor of 1.6 more. And besides that, the belief that in animals cognition is working in different ways than in humans, seems to be an error; researchers found you can talk with chimpanzees if you simply learn their language, and they can learn ours. Capra summarizes this research shortly:

The unified, post-Cartesian view of mind, matter, and life also implies a radical reassessment of the relationships between humans and animals. Throughout most of Western philosophy, the capacity to reason was seen as a uniquely human characteristic, distinguishing us from all other animals. The communication studies with chimpanzees have exposed the fallacy of this belief in the most dramatic of ways. They make it clear that the cognitive and emotional lives of animals and humans differ only by degree; that life is a great continuum in which differences between species are gradual and evolutionary. (Id., 65-66).

The Social Critic

I shall finalize this chapter with pointing to some *political* hidden connections that Capra unveils in his book. There are probably still people around who are fond of *biotechnology*, but I guess they just ignore the facts, and their knowledge may be the result of disinformation.

> The most widespread use of plant biotechnology has been to develop herbicide-tolerant crops in order to boast the sales of particular herbicides. There is a strong likelihood that the transgenic plants will cross-pollinate with wild relatives in their surroundings, thus creating herbicide-resistant superweeds. Evidence indicates that such gene flows between transgenic crops and wild relatives are already occurring. (Id., 193)

Why do we need biotechnology, if I may ask? And—is it democratic to favor legal and social policies that bring damage to our planet? I learnt as a law student that such a system is called an *oligarchy*, the reign of a few who control the rest. So how did we ever come to say that we are living in a *democracy*?

> In the animal kingdom, where cellular complexity is much higher, the side effects in genetically modified species are much worse. 'Super-salmon' which were engineered to grow as fast as possible, ended up with monstrous heads and died from not being able to breathe or feed properly. Similarly, a superpig with a human gene for a growth hormone turned out ulcerous, blind, and impotent. (...) The most horrifying and by now best-known story is probably that of the genetically altered hormone called recombinant bovine growth hormone, which has been used to stimulate milk production in cows despite the fact that American dairy farmers have produced vastly more milk than people can consume for the past fifty years. The effects of this genetic engineering folly on the cow's health are serious. They include bloat, diarrhea, diseases of the knees and feet, cystic ovaries, and many more. Besides, their milk may contain a substance that has been implicated in human breast and stomach cancers. (Id., 198)

Why do we need superpigs? It seems to me that they are the result of quantitative thinking, a primacy of quantity over quality, and this for the obvious reason of maximizing profits. This is a good example for the fact that we live in what has been called the 'corporate society' or *Corporate America*, as the prototype of a society in which major corporations dictate the standards the government is going to follow and to enact as laws. Capra notes the details:

In the United States, the biotech industry has persuaded the Food and Drug Administration (FDA) to treat GM food as substantially equivalent to traditional food, which allows food producers to evade normal testing by the FDA and the Environmental Protection Agency (EPA), and also leaves it to the companies' own discretion as to whether to label their products as genetically modified. Thus, the public is kept unaware of the rapid spread of transgenic foods and scientists will find it much harder to trace harmful effects. Indeed, buying organic is now the only way to avoid GM foods. (Id., 199)

In Germany and France, the laws are different regarding genetically modified food and the EU will probably ban all products that are to be subsumed under this term, because this is already the state of the law in Germany and France, and for good reasons. Capra informs:

The governments of France, Italy, Greece, and Denmark announced that they would block the approval of new GM crops in the European Union. The European Commission made the labeling of GM foods mandatory, as did the governments of Japan, South Korea, Australia, and Mexico. In January 2000, 130 nations signed the groundbreaking Cartagena Protocol on Biosafety in Montreal, which gives nations the right to refuse entry to any genetically modified forms of life, despite vehement opposition from the United States. (Id., 228)

As a lawyer, I can clearly see that we are facing currently a challenge to legally codify these new technologies —or they as it were are going to codify *us*, entraining us in a turbulence of faits établis, and the law will leap behind the actual developments. But the law should better accompany the research step by step so as to be informed by the *explosive growth* of these heavily funded research disciplines. Capra writes:

> The development of such new biotechnologies will be a tremendous intellectual challenge, because we still do not understand how nature developed technologies during billions of years of evolution that are far superior to our human designs. How do mussels produce glue that sticks to anything in water? How do spiders spin a silk thread that, ounce for ounce, is five times stronger than steel? How do abalone grow a shell that is twice as tough as our high-tech ceramics? How do these creatures manufacture their miracle materials in water, at room temperature, silently, and without any toxic byproducts? (Id., 204)

I guess our intellectual and political challenge for the next decade is to take responsibility and stop the development of these technologies altogether. This can only be done through precise legal codification, where *rule and exception* are clearly stated. The *rule* should be the total ban of biotechnology, and the *exceptions* should be the limited cases where their use can be responsibly allowed without jeopardizing living systems.

We should eventually begin to understand that our intelligence simply is not on the same level as nature's, and playing around with that kind of stuff is not child play.

In providing this information to the reader, Capra proves to be a critical citizen, and in this quality he is more than a scientist. That Capra also can be a pragmatist, I am going to show in the last sub-chapter, discussing one of his latest books, which is a sampler, co-edited with Wolfgang Pauli, and published by the United Nations.

The Pragmatist

Steering Business Toward Sustainability (1995) is a book of high practical value for leaders and organizations who are *conscious of the need for deep ecology* and the challenge we presently face to update most of our basic business and investment routines and procedures in order to build sustainable organizations.

Capra speaks true language in this rather pragmatic book that is different in quality than his former publications in that it addresses business questions, and in that it presents the insights of a consultant who is often around NGOs and government, knowing how slowly things move in real life. Contrary to many others in these professions, Capra keeps a sober and realistic perspective right at the onset of this booklet:

> Quite simply, our business practices are destroying life on earth. Given current corporate practices, not one wildlife reserve, wilderness, or indigenous culture will survive the global market economy. (Id., 1)

Capra's idea of *ecology* has grown over many years. It is rooted in the insights he exposed in his previous four books; this means that his ecological insights and ideas are solidly grounded.

In addition, Capra leaves no doubt that *ecological literacy* is not just a new science concept, but *an intrinsically spiritual idea*. He also credits religions and peoples who have practiced ecological thinking long before the birth of the United States of America:

> When the concept of the human spirit is understood as the mode of consciousness in which the individual feels connected to the cosmos as a whole, it becomes clear that ecological awareness is spiritual in its deepest essence. It is therefore not surprising that the emerging new vision of reality, based on deep ecological awareness, is consistent with the so-called perennial philosophy of spiritual traditions, whether we talk about the spirituality of Christian mystics, that of Buddhists, or the philosophy and cosmology underlying the American Indian traditions. (Id., 3)

Capra reminds us of the fact that when restructuring our economies, we should learn from nature, instead of feeling superior over nature. *Ecological literacy* is one of the notions Capra is currently lecturing about, and Gunter Pauli, the co-editor of this reader is one of Capra's truest collaborators, and himself an authority on ecology in Germany.

—See also the Web Presence of Ecological Literacy or Ecoliteracy: http://www.ecoliteracy.org/

Within the concept of ecological literacy, Capra seems to give highest importance to the term *sustainability*, and comprehensively explains what this term means:

> In our attempts to build and nurture sustainable communities we can learn valuable lessons from ecosystems, because ecosystems are sustainable communities of plants, animals, and microorganisms. To understand these lessons, we need to learn nature's language. We need to become ecologically literate. (…) Being ecologically literate means understanding how ecosystems organize themselves so as to maximize sustainability. (Id., 4)

Many of us have not yet understood why technologies are so much in conflict with nature's setup, and this is a fact that is hardly ever elucidated in the mass media. Non-educated people, and even entrepreneurs who have not been exposed to academic study are usually at pains with understanding the deeper reasons of this conflict. Capra, referencing Paul Hawken, *The Ecology of Commerce*, elucidates it:

> The present clash between business and nature, between economics and ecology, is mainly due to the fact that nature is cyclical, whereas our industrial systems are linear, taking up energy and resources from the earth, transforming them into products plus waste, discarding the waste, and finally throwing away also the products after they have been used. Sustainable patterns of production and consumption need to be cyclical, imitating the processes in ecosystems. (Id., 5)

Back in Antiquity, there was hardly a need for people to learn systems thinking because they were *naturally aligned with the logic of nature;* they were simply living *with* nature, and not on top of nature, as we do today. We can also say that we as modern city dwellers have lost our *continuum*, as it was expressed with much emphasis by Jean Liedloff in *The Continuum Concept (1977)*. Besides, Capra informs about how we should apply ecology in our daily lives, and what it teaches us. There are seven principles to learn that Capra calls *Principles of Ecology* and that he explains one by one:

Interdependence
All members of an ecosystem are interconnected in a web of relationships, in which all life processes depend on one another.

Ecological Cycles
The interdependencies among the members of an ecosystem involve the exchange of energy and resources in continual cycles.

Energy Flow
Solar Energy, transformed into chemical energy by the photosynthesis of green plants, drives all ecological cycles.

Partnership
All living members of an ecosystem are engaged in a subtle interplay of competition and cooperation, involving countless forms of partnership.

Flexibility
Ecological cycles have the tendency to maintain themselves in a flexible state, characterized by interdependent fluctuations of their variables.

Diversity
The stability of an ecosystem depends on the degree of complexity of its network of relationships; in other words, on the diversity of the ecosystem.

Coevolution
Most species in an ecosystem coevolve through an interplay of creation and mutual adaptation.

Sustainability
The long-term survival of each species in an ecosystem depends on a limited resource base. Ecosystems organize

themselves according to the principles summarized above so as to maintain sustainability. (Id., 6)

Capra also elucidates the *feedback-looping* that we find is a typical feature of living systems. The understanding of *feedbacking by constant parameter change* as a response to a given stimulus is crucial for the understanding of the cyclic nature of all life. This is one of the points modern scientists are really at pains with because their thought structure is too linear. Capra explains:

> When changing environmental conditions disturb one link in an ecological cycle, the entire cycle acts as a self-regulating feedback loop and soon brings the situation back into balance. And since these disturbances happen all the time, the variables in an ecological cycle fluctuate continually. These fluctuations represent the ecosystem's flexibility. Lack of flexibility manifests itself as stress. In particular, stress will occur when one or more variables of the system are pushed to their extreme values, which induces increased rigidity throughout the system. Temporary stress is an essential aspect of life, but prolonged stress is harmful and destructive to the system. (Id., 7)

It's exactly this *widely unpredictable feedback-looping* that is at the root of the current worldwide ecological destruction and, hopefully, regeneration. This dangerous situation is worsened by the general lack of ecological literacy about the possible effects of large disturbances, such as ozone hole, deforestation, global warming and desertification.

Our knowledge also is insufficient to make ecological solutions work effectively even once ecology-friendly policies are implemented by governments and companies. It is not enough to see the dangers and implement good new laws for protecting nature because we also need to understand how the damages that are already done will interact with our new policies; this is so because it's not taken for granted that our best-intended tactics of healing nature are *really* healing nature. For insuring this, we have to learn much more about feedback-looping in natural systems.

We must understand that nature heals herself and that we only need to remove the factors that cause the damage. For example, it has been shown that the planting of new trees does not per see heal the damage that deforestation has done to ecological saneness of our planet. It's all in the *why and how* of planting trees, where, how many, and in what mixture of species that the wisdom lies. On the other hand, it has been seen in Indonesia, which is one of the worst hit countries by deforestation, that huge areas that were deforested began to grow trees *without anybody doing anything about it!* Later research showed that the conditions had been ideal for trees to grow again, but *nobody really knew* why at other places, where at first sight conditions were very similar, this was not the case.

We definitely have to practice *humility* in the face of the dreadful ignorance we are suffering from regarding the complexity level of nature, at all phases of evolution.

We are simply not trained in complexity thinking, and our schools and universities destroy the little of complexity we have developed naturally as children as a result of free play. It is *freedom* that is at the basis of building complexity, not discipline, it is *permissiveness*, not repression.

Here is where our *moralism* clearly stares grimly in nature's face because nature is immoral or moral-neutral. As an insight, we should do away with our projections upon nature and at the same time get all our senses and our emotional intelligence ready for *receiving the messages of nature*. Nature communicates when we are ready to listen, and it will tell us how we can help healing the damage we have done over five thousand years of patriarchal ignorance.

The present book together with *The Hidden Connections* and *The Web of Life* teaches the basics of understanding nature's complexity and the importance of *diversity*, a concept that at present is rather shunned by mainstream politics, while liberal phases, as it was the case through the 1970s, foster higher levels of cultural diversity.

Nature shows us that this is not just a random development but that it's diversity on which side is intelligent and life-fostering behavior, and not uniformity. This is so, inter alia, because diversity fosters *flexibility*, and vice versa, uniformity entails rigidity. Capra elucidates:

> In ecosystems, flexibility through fluctuations does not always work, because there can be very severe disturbances that actually wipe out an entire species. In other words, one

of the links in the ecosystem's network is destroyed. An ecological community will be resilient when this link is not the only one of its kind; when there are other connections that can at least partially fulfill its functions. In other words, the more complex the network, the greater the diversity of its interconnections, the more resilient it will be. The same is true in human communities. Diversity means many different relationships, many different approaches to the same problem. A diverse community is a resilient community, capable of adapting easily to changing situations. (Id., 8)

What does *loss of diversity* on the planet, in all layers of living systems, mean for our future? The regard is rather dim, and Capra leaves no doubt about this:

The loss of biodiversity, i.e. the daily loss of species, is in the long run one of our most severe global environmental problems. And because of the close integration of tribal indigenous people into their ecosystems, the loss of biodiversity is closely tied to the loss of cultural diversity, the extinction of traditional tribal cultures. This is especially important today. As the beliefs and practices of the industrial culture are being recognized as part of the global ecological crisis, there is an urgent need for a wider understanding of cultural patterns that are sustainable. The vast folk wisdom of American Indian, African, and Asian traditions has been viewed as inferior and backward by the industrial culture. It is time to reverse this Euro-centric arrogance and to recognize that many of these traditions—their ways of knowing, technologies, knowledge of foods and medicines, forms of aesthetic expression, patterns of social interaction, communal relationships, etc.—embody the ecological wisdom we so urgently need today. (Id.)

This is what I am saying since more than 15 years, having founded, back in 1994, in Rotterdam, Holland, *Ayuda Foundation International* for the protection of tribal people's life wisdom and their high cultural diversity, and wistful traditions for the healing and integration of emotions.

It's a fact that in most developing countries technologies for recycling and for healing our badly afflicted ecosystems are costly and not as accessible and readily available as in high-tech nations. Only truly supportive cultural and technological exchange between rich and poor countries can help changing this dim picture.

Whatever our personal opinions are in the face of these huge global problems that our next generations will inevitably be burdened with, we have to keep an open mind and learn, and change our rigid positions.

Fritjof Capra and Wolfgang Pauli have given in this reader very useful suggestions that can be taken as starting points for deeper study, as the field of investigation is huge, and complicated.

Nature's complexity is perhaps the *single most important topic* of study for 21st century science, and I hope I can contribute a little to it with my own publishing efforts. As for the authors of this book, they surely have done their very substantial contributions!

Books Reviewed

The Tao of Physics (1975)
Green Politics (1986)
The Turning Point (1987)
Uncommon Wisdom (1989)
Belonging to the Universe (1991)
Steering Business Toward Sustainability (1995)
The Web of Life (1997)
The Hidden Connections (2002)
The Science of Leonardo (2007)
Learning from Leonardo (2013)
The Systems View of Life (2014)

Chapter Three
The Tao of Physics

The Tao of Physics

An Exploration of the Parallels between Modern Physics and Eastern Mysticism

New York: Shambhala, 1975
With original graphics

Review

My book reviews in this volume do not simply repeat what was pointed out in Chapter Two. Fact is that I have read all Capra books twice. In the case of this present book I have first read the revised Bantam Books 1984 edition and quoted upon it in Chapter Two. Here in my review, I will use quotes I have taken from the original 1975 Shambhala edition.

There is an interesting phenomenon when you re-read an important book many years after you read it for the first time. You will discover things and focus upon passages different from those you discovered and found noteworthy first. I was reading the *Tao of Physics* first in 1986 and the quotes in Chapter Two are from that first lecture. Then, in 2014 I was reading the book again, and preferred the original edition of it, as it is much better edited, and comes with a much better layout, font setting and presentation. This may be a bit of a subjective impression, however myself not just being the author but also the designer and publisher of my books, I react rather sensitively to how books are crafted, how they are designed and present

CHAPTER THREE

themselves to the reader. I found over the years many Bantam books negligently edited, even containing typos, while Shambhala brings out books that are perfect in every respect, because the design, font settings, layout and presentation perfectly match the nature of the content. This is so much the more the case as some of the graphics in this edition are Capra's own ones, and they are missing in the Bantam edition.

I would like to start my discussion of the book with the first chapter: 'Modern Physics, a Path with a Heart?' It is a very good title for it explains why physics is so important in our times, but it also questions it along the lines of this quote from the books of Carlos Castaneda:

> Any path is only a path, and there is no affront, to oneself or to others, in dropping it if that is what your heart tells you … Look at every path closely and deliberately. Try it as many times as you think necessary. Then ask yourself, and yourself alone, one question … Does this path have a heart? If it does, the path is good; if it doesn't it is of no use.
> —Carlos Castaneda, *The Teachings of Don Juan*

Fritjof Capra starts with explaining that atomic physics has a high influence in our world, in politics, in weaponry, and extends to thought and culture. During the 20th century, then, a radical revision of classical concepts of physics took place, as the concept of matter in subatomic physics is totally different from the traditional idea of material substance. The same, he says, is true with regard to concepts like space, time, or cause and effect. He then clarifies that the purpose of the book is to explore the relationship between the concepts of modern physics and the basic ideas in the philosophical and religious traditions of the Far East.

He also clarifies upfront that when he refers to 'Eastern mysticism,' he means the religious philosophies of Hinduism, Buddhism and Taoism. He then writes:

> Although these comprise a vast number of subtly interwoven spiritual disciplines and philosophical systems, the basic features of their world view are the same. This view is not limited to the East, but can be found to some degree in all mystically oriented philosophies. The argument of this book could therefore be phrased more generally, by saying that modern physics leads us to a view of the world which is very similar to the views held by mystics of all ages and traditions. (…) I shall therefore, for the sake of simplicity, talk about the 'Eastern world view' and shall only occasionally mention other sources of mystical thought.' (Id., 19)

The argument the author makes at the starting point of his book, as a hypothesis to be argued through to the end, is that the pathway of Western science was basically cyclic. It started as a mystical tradition about 2500 years ago with the mystical philosophies of the early Greeks, 'rising and unfolding in an impressive development of intellectual thought that increasingly turned away form its mystical origins to develop a world view which is in sharp contrast to that of the Far East.' (Id.) Then, in our time, we are

CHAPTER THREE

finally coming back to the early Greek and Eastern philosophies.

> This time, however, it is not only based on intuition, but also on experiments of great precision and sophistication, and on a rigorous and consistent mathematical formalism. (Id.)

While the reader may want to contradict to this rather daring hypothesis that sounds too simplistic to be true, the author reasons his case through with quite some well-reflected arguments; in fact the rest of this chapter consists of precisely 12 steps that retrace the development of Western science, and as such is quite an intellectual *tour de force,* and a major achievement of synthetic thought! I adapt this section for my review here without quoting. My summary extends from page 20 to page 25 of the book—

1) We are starting our journey in the 6th century B.C. in Greece, at the time being a culture where science, philosophy and religion were not separated. The reigning philosophical school were the Milesians in Ionia. They had an interesting idea about nature, calling it 'physis'—from which was derived 'physics'—which meant something so broad-minded as the essential nature of all things. These sages were thus holistic thinkers of the first order. Therefore, they were called 'hylozoists' which means something like 'those who think that matter is alive;' they made no distinction between animate and inanimate nature. The two main philosophers of this school were Thales and Anaximander. Thales declared all things to be full of gods and Anaximander explained the living nature

of things with the notion of 'pneuma' or cosmic breath which was one of the first words describing the vital energy or what Barbara Brennan coined the universal energy field (UEF).

2) The organic worldview of the Milesians finds a parallel in ancient Indian and Chinese philosophy. There is one towering name in Greek philosophy who can be said to have the most closely reflected the Eastern view of the world: *Heraclitus of Ephesus.* Just as the *I Ching* or *Book of Changes* in China, Heraclitus explained nature as continual change and transformation. He taught a theory very similar to the Chinese notion of *Yin* and *Yang* in that he emphasized the dynamic and cyclic interplay of opposites as the driving motor of life and growth. This unity, which contains and transcends all opposing forces, he called the *Logos*.

3) The split of this holistic and organic worldview began with the Eleatic school, which posited a 'Divine Principle' standing above nature. This principle developed into the predominant idea of a personal God who stands above the world and directs it. This ultimately led then to the schizoid split between mind and matter that became so characteristic of Western science and philosophy.

4) This split was emphasized by Parmenides who declared himself in opposition to Heraclitus. He went as far as saying that to see change in nature was an illusion of the senses.

5) A hundred years later attempts were made in Greece to overcome the split and reconcile Parmenides' and Heraclitus' ideas. This led to the idea of the *atom* as the smallest and indivisible unit of matter. The main exponents of this philosophical school were Leucippus and Democritus. The further idea then was that there are 'basic building blocks' in nature. However, they were not seen as organic, but essentially dead matter.

6) Logically, with splitting off mind and matter, from that time philosophers devoted their thinking to the spiritual world and problems of ethics, thereby more and more emphasizing the view that body and soul are two different realms and are to be looked at by different scientific or religious doctrines.

7) With the towering figure of Aristotle, the scientific knowledge of Antiquity was systematized and organized and the resulting worldview was cemented for 2000 years to come as the Christian Church supported Aristotle's doctrine all through the Middle Ages.

8) In the late 15th century the study of nature was for the first time approached with what later became called 'the scientific method,' that is, the undertaking of scientific experiments that are expressed in mathematical language as pioneered by Galileo.

9) The most extreme form of the spirit/matter dualism was then conceptualized and formalized by Descartes in France who based his view of nature on a division between *res cogitans* (mind) and *res extensa* (matter). This 'Cartesian'

setup of science led straight to a worldview in which the world was seen as a gigantic clockwork.

10) The mechanistic view of the world was then formulated scientifically by Isaac Newton; it was then made the foundation of classical physics, also called 'mechanics.'

a) From the second half of the 17th century to the end of the 19th century, the mechanistic model of the universe was reigning in science, paralleled by a *God King* who ruled the world from above; the fundamental laws of nature consequently became declared as the laws of God to which the world was subjected.

b) Descartes' idea 'cogito ergo sum'—I think, therefore I exist—led to the ultimate concept of human beings as isolated egos existing 'inside' their bodies. As a result, the ancient organic view of the body got lost with this idea of a split in each individual leading to a large number of separate compartments.

c) The *fragmented view* of nature and all living was then paralleled in society by the split of the political landscape in different nations, races, religious and political groups, thereby creating the perceptual disturbance that led to the acute crisis of perception we are presently dealing with on a worldwide scale. It has brought about a grossly unjust distribution of natural resources creating disorder and violence, as well as environmental destruction and the loss of many species.

CHAPTER THREE

11) In contrast to the mechanistic worldview reigning in the West, the Eastern worldview remained organic. As expressed in Buddhist philosophy—

> When the mind is disturbed, the multiplicity of things is produced, but when the mind is quieted, the multiplicity of things disappears.

12) While in Eastern mysticism, be it *Hinduism, Buddhism* or *Taoism*, there are various schools, they do have in common that they all emphasize the basic unity of the universe and see the 'ten thousand things' as basically interrelated. It's for this reason that these philosophies may be called 'religious' in nature. They also share the idea that all in nature is ever-changing and that *human consciousness is forever fluid*. The cosmos is seen as alive, in motion, and organic, both spiritual and material at the same time.

It is from this point onwards that Fritjof Capra sets out to compare the Eastern worldview with the fragmented Western concepts of science on one hand, and religion, on the other. He writes:

> This book aims at improving the image of science by showing that there is an essential harmony between the spirit of Eastern wisdom and Western science. It attempts to suggest that modern physics goes far beyond technology, that the way—or Tao—of physics can be a path with a heart, a way to spiritual knowledge and self-realization. /25

It is highly intriguing to see after this review of the development of our science paradigm that we indeed seem to return to the very origins, namely the organic and holistic view of life, and of nature that was intuitively formulated by Thales, Anaximander, and especially Heraclitus. In the meantime, and going beyond the books of Fritjof Capra, the cosmic energy field and the human energy field are scientifically recognized. We speak about the Field, the Quantum Field, the Zero Point Field or the Quantum Vacuum. It is recognized today that while the idea of an 'ether' was not maintained for, as Albert Einstein found, the field concept itself is dynamic and conveys the conduit function of the ether, we do not need, as Rupert Sheldrakes writes in *A New Science of Life (1995)* any 'vitalistic' theories to explain what he himself coined the 'morphogenetic field.' Capra writes:

> It was Einstein who clearly recognized this fact fifty years later when he declared that no ether existed and that the electromagnetic fields were physical entities in their own right which could travel through empty space and could not be explained mechanically. /61

This is after all a remarkable epistemological progress compared to the old concept of the ether which was after all seen as an add-on to the four elements, water, wind, fire and earth. But from a conceptual point of view it doesn't make sense to see the ether, the vital energy field, as separate from the other elements. In fact, this field is contained in them, as it is contained in all, it's universal and nonlocal, and therefore, ubiquitous.

In contradistinction to the very beginnings of science in ancient Greece, today we are equipped with highly precise machinery and measurement devices for locating and manipulating the field, and we can setup a multitude of experiments and computerized research projects in order to prove our assumptions right, or to falsify them. Fritjof Capra writes:

> Now the force concept was replaced by the much subtler concept of a field which had its own reality and could be studied without any reference to material bodies. /60-61

The next point where we see today the error of the 'atomistic' Greek philosophers Leucippus and Democritus is the discovery, made first by Rutherford and then was confirmed by others, that the atom resembles a mini-galaxy and has nothing about it that can incite us to think it was a 'hard' and 'dead' material substance. Capra writes:

> When Rutherford bombarded atoms with ... alpha particles, he obtained sensational and totally unexpected results. Far from being the hard and solid particles they were believed to be since antiquity, the atoms turned out to consist of vast regions of space in which extremely small particles—the electrons—moved around the nucleus, bound to it by electric forces. /65

The next point in Capra's scheme of argumentation is to show how *relativity theory* prepared the discovery of the quantum field in showing that space and time are fully equivalent:

> Space and time are fully equivalent; they are unified into a four-dimensional continuum in which the particle

interactions can stretch in any direction. If we want to picture these interactions, we have to picture them in one 'four-dimensional snap shot' covering the whole span of time as well as the whole region of space. /185

What really is meant by Einstein's space-time concept? J. Krishnamurti and other Eastern teachers emphasize that thought must take place in time. However, Capra explains that space-time of relativistic physics is a 'similar timeless space of higher dimensions' in which all events are interconnected, but that these connections are not causal.

> Particle interactions can be interpreted in terms of cause and effect only when the space-time diagrams are read in a definite direction, e.g. from the bottom to the top. When they are taken as four-dimensional patterns without any definite direction of time attached to them, there is no 'before' and no 'after,' and thus no causation. /186

In fact, modern physics has transcended the materialistic worldview in that it pictures matter not as passive and inert, but as being in a vibrational continuum, a kind of dance 'whose rhythmic patterns are determined by the molecular, atomic and nuclear structures.' (Id., 194)

In addition, quantum theory has shown that particles are probability patterns, interconnections in an inseparable cosmic web.

> The particles of the subatomic world are not only active in the sense of moving around very fast; they themselves are processes! The existence of matter and its activity cannot be

separated. They are but different aspects of the same space-time reality. /203

Now, how does this compare to the Buddhist worldview? Capra writes:

> Like modern physicists, Buddhists see all objects as processes in a universal flux and deny the existence of any material substance. This denial is one of the most characteristic features of all schools of Buddhist philosophy. It is also characteristic of Chinese thought which developed a similar view of things as transitory stages in the ever-flowing Tao and was more concerned with their interrelations than with their reduction to a fundamental substance. 'While European philosophy tended to find reality in substance,' writes Joseph Needham, 'Chinese philosophy tended to find it in relation.' /204

Let us remember that matter and empty space were two fundamentally distinct concepts on which the atomism of Democritus and Newton was based. However, in relativity theory these two concepts are no more separated.

At the subatomic level, things become even more volatile where classical field theory and quantum theory need to be combined to describe the interactions between subatomic particles.

Critique

After the praise I have for all of Capra's books, I would like to inform the reader that *The Tao* has been received with a mixed sort of enthusiasm; among scientists critical voices were at times explicit. It is important to understand both the value of the book and the value of some of the critique that was uttered.

Before pointing this out more in detail, I would myself plead for not bringing young science students too early in contact with this book, for the understanding of Capra's basic message is difficult to apprehend for children as they are not familiar with mysticism. This is not because of cultural differences between East and West, but because mysticism is something only a mature mind can comprehend.

Living myself in South-East Asia since more than 20 years, I can say with conviction that children in Asia are in as much unfamiliar with their own mystical tradition as Western children are with ours—or theirs, for that matter.

It requires a high level of abstract thinking ability to set out to compare science with mysticism, no matter which kind of traditions we are talking about. Science as the empirical search for truth, through experimentation and the falsification of theories seems to have rather little in common with mystical thinking in its rather intuitive and non-empirical focus upon reality. While it is certainly true that even the most rational-minded and empirical scientist uses intuition for progressing in their research—for that fact Albert Einstein is certainly a prime example—it is not for that matter taken for granted to compare an entire science tradition with an entire mystical tradition, so much the more both traditions have grown in very different cultural soil.

This quotation may show how problematic this point of departure can be, as the author actually unveils the conceptual trap:

> [The] reality of the Eastern mystic cannot be identified with the quantum field of the physicist because it is seen as the essence of *all* phenomena in this world and, consequently, is beyond all concepts and ideas.
>
> The quantum field, on the other hand, is a well-defined concept which only accounts for some of the physical phenomena.
>
> Nevertheless, the intuition behind the physicist's interpretation of the subatomic world, in terms of the quantum field, is closely paralleled by that of the Eastern mystic who interprets his or her experience of the world in terms of an ultimate underlying reality. Subsequent to the emergence of the field concept, physicists have attempted to unify the various fields into a single fundamental field which would incorporate all physical phenomena. Einstein, in particular, spent the last years of his life searching for such a unified field.
>
> The *Brahman* of the Hindus, like the *Dharmakaya* of the Buddhists and the *Tao* of the Taoists, can be seen, perhaps, as the ultimate unified field from which spring not only the phenomena studied in physics, but all other phenomena as well. /211

This is certainly well put, but it is what it is: a proposition that cannot be verified by science. It's a philosophical statement and I cannot see how it could in any way serve as a pointer to scientific reality? Capra continues:

> The Taoists ascribe a similar infinite and endless creativity to the *Tao* and, once again, call it empty. 'The Tao of Heaven is empty and formless' says the Kuan-tzu, and Lao Tzu uses several metaphors to illustrate this emptiness. He often compares the *Tao* to a hollow valley, or to a vessel which is forever empty and thus has the potential of containing an infinity of things. /212

While Capra makes his point thoroughly and with very good arguments, the original thesis is a daring one, and from an epistemological point of view, it's unusual to find a science tradition criticized with other than scientific arguments. The following quote shows clearly that the concept of the quantum field as a 'standalone definition' that can very well survive without being informed by Eastern mysticism:

> With the concept of the quantum field, modern physics has found an unexpected answer to the old question of whether matter consists of indivisible atoms or of an underlying continuum. The field is a continuum which is present everywhere in space and yet its particle aspect has a discontinuous, 'granular' structure. The two apparently contradictory concepts are thus unified and seen to be merely different aspects of the same reality. /215

In my youthful enthusiasm when reading *The Tao* thirty years ago, I did not question these assumptions, but today I am wondering about some of Capra's tenets, for example his almost obsessive affirmation of female values as being superior to male values, or *yin* being more valuable than *yang*.

But even if one accepts that our culture is biased in this respect, the cultures of the East have their share in the domination of the female by the male, and the large-scale destruction of nature. I have not found a single streak of ecological thinking in Asia in all those twenty years living and working here; virtually *nobody cares for the trees* here, not to mention the rest of nature; trees are just slaughtered at convenience and the timber serves to fire stoves or even is sold to neighboring countries.

While governments are at a tender beginning to forge laws that serve to protect nature, nobody seems to enforce those laws, and their tenor of 'sustainability' does not seem to enter the mentality of the people. In the face of this reality it seems like an irony to take the mystical traditions of these cultures to criticize our own science!

This being said, I believe that *The Tao* had to appear at the time it was published, as today the typical new-ageism of that time (the 1970s) is less appealing to people, if they do not see the whole trend as just another fashion or another fad that will eventually be outlived and forgotten.

This next quote shows again that our modern science has developed the mechanism for dealing with reality in a systemic matter, without needing epistemological aid and guidance from wisdom traditions of the past:

> The exploration of the subatomic world in the twentieth century has revealed the intrinsically dynamic nature of matter. It has shown that the constituents of atoms, the subatomic particles, are dynamic patterns which do not exist as isolated entities, but as integral parts of an inseparable network of interactions.
>
> These interactions involve a ceaseless flow of energy manifesting itself as the exchange of particles; a dynamic interplay in which particles are created and destroyed without end in a continual variation of energy patterns. The particle interactions give rise to the stable structures which build up the material world, which again do not remain static, but oscillate in rhythmic movements. The whole universe is thus engaged in endless motion and activity; in a continual cosmic dance of energy. /225

After this personal remark, I would like to shortly outline the criticism of a study published in Germany by a German scientist, in 1993. The translated title of the book is *The New Mysticism. Eastern Mysticism and Modern Natural Science in New Age Thinking (Würzburg, 1993).* The author H.G. Russ, criticizes Capra's opinion modern science was founded upon a belief system. He positively discusses Ken Wilber's idea that physics and mysticism are not different ways leading to the same reality, but ways that lead to different levels of reality.

As Capra speaks as a physicist and discusses physics in the framework of mysticism, the author forwards arguments demonstrating that Capra's epistemological stance is not a strong one. In addition, he argues that

CHAPTER THREE

Capra has not recognized the difference between scientific holism and holism at the basis of mystical thinking. The next quote really shows a mistaken approach:

> The principal schools of Eastern mysticism thus agree with the view of the bootstrap philosophy ... /292

Do schools of mysticism need to agree with a modern science theory known as 'bootstrap?' The very idea is ridiculous in the first place!

While mystics typically experienced reality in a holistic manner, this was not the case, and could not be the case for scientists.

He further points out that systems theory was not per se a scientific approach that could be qualified as 'holistic.' Finally, he argues that as science did not contain any religious elements, any message that borders religion could never be accurately derived from scientific inquiry.

I find these arguments valid enough for being scrutinized further; they can in my view not easily be wiped off the table. It is therefore recommended to see *The Tao* as an intuitive and well positively engaging attempt to bring intuition into the scientific method, as a balancing factor, but for that matter I would be *cautious to systematically deconstruct modern science* with the arguments forwarded by Capra in this book.

Quotes

- Modern physics has had a profound influence on almost all aspects of human society. It has become the basis of natural science, and the combination of natural and technical science has fundamentally changed the conditions of life on our earth, both in beneficial and detrimental ways. Today, there is hardly an industry that does not make use of the results of atomic physics, and the influence these have had on the political structure of the world through their application to atomic weaponry is well known. However, the influence of modern physics goes beyond technology. It extends to the realm of thought and culture where it has led to a deep revision in our conception of the universe and of our relation to it. /3

- This view is not limited to the East, but can be found to some degree in all mystically oriented philosophies. The argument of this book could therefore be phrased more generally by saying that modern physics leads us to a view of the world which is very similar to the views held by mystics of all ages and traditions. Mystical traditions are present in all religions, and mystical elements can be found in many schools of Western philosophy. The parallels to modern physics appear not only in the Vedas of Hinduism, in the I Ching, or in the Buddhist sutras, but also in the fragments of Heraclitus, in the Sufism of Ibn Arabi, or in the teachings of the Yaqui sorcerer Don Juan. The difference between Eastern and Western mysticism is that mystical schools have always played a marginal role in the West, whereas they constitute the mainstream of Eastern philosophical and religious thought. /5

- If physics leads us today to a world view which is essentially mystical, it returns, in a way, to its beginning, 2500 years ago. It is interesting to follow the evolution of Western science along its spiral path, starting from the mystical philosophies of the early Greeks, rising and / unfolding in an impressive development of intellectual thought that increasingly turned away from its mystical origins to develop a world view which is in sharp contrast to that of the Far East. In its most recent stages, Western science is finally overcoming this view and coming back to those of the early Greek and the Eastern philosophies. This time, however, it is not only based on intuition, but also on

CHAPTER THREE

experiments of great precision and sophistication, and on a rigorous and consistent mathematical formalism. /5-6

> The roots of physics, as of all Western science, are to be found in the first period of Greek philosophy in the sixth century B.C., in a culture where science, philosophy and religion were not separated. The sages of the Milesian school in Ionia were not concerned with such distinctions. Their aim was to discover the essential nature, or real constitution, of things which they called 'physis.' The term 'physics' is derived from this Greek word and meant therefore, originally, the endeavor of seeing the essential nature of all things.
>
> This, of course, is also the central aim of all mystics, and the philosophy of the Milesian school did indeed have a strong mystical flavor. The Milesians were called 'hylozoists', or 'those who think matter is alive,' by the later Greeks, because they saw no distinction between animate and inanimate, spirit and matter. In fact, they did not even have a word for matter, since they saw all forms of existence as manifestations of the 'physis,' endowed with life and spirituality. Thus Thales declared all things to be full of gods and Anaximander saw the universe as a kind of organism which was supported by 'pneuma,' the cosmic breath, in the same way as the human body is supported by air.
>
> The monistic and organic view of the Milesians was very close to that of ancient Indian and Chinese philosophy, and the parallels to Eastern thought are even stronger in the philosophy of Heraclitus of Ephesus. Heraclitus believed in a world of perpetual change, of eternal 'Becoming.' For him, all static Being was based on deception, and his universal principle was fire, a symbol for the continuous flow and change of all things. Heraclitus taught that all changes in the world arise from the dynamic and cyclic interplay of opposites, and / he saw any pair of opposites as unity. This unity, which contains and transcends all opposing forces, he called the Logos. /6-7

> The split of this unity began with the Eleatic school, which assumed a Divine Principle standing above all gods and men. This principle was first identified with the unity of the universe, but was later seen as an intelligent and personal God who stands above the world and directs it. Thus began a trend of thought which led, ultimately, to the separation of spirit and matter and to a dualism which became characteristic of Western philosophy. /7

- A drastic step in this direction was taken by Parmenides of Elea, who was in strong opposition to Heraclitus. (...) This led to the concept of the atom, the smallest indivisible unit of matter, which found its clearest expression in the philosophy of Leucippus and Democritus. /7

- As the idea of a division between spirit and matter took hold, the philosophers turned their attention to the spiritual world, rather than the material, to the human / soul and the problems of ethics. These questions were to occupy Western thought for more than two thousand years after the culmination of Greek science and culture in the fifth and fourth centuries B.C. /6-7

- The scientific knowledge of antiquity was systematized and organized by Aristotle who created the scheme which was to be the basis of the Western view of the universe for two thousand years. But Aristotle himself believed that questions concerning the human soul and the contemplation of God's perfection were much more valuable than investigations of the material world. The reason the Aristotelian model of the universe remained unchallenged for so long was precisely this lack of interest in the material world, and the strong hold of the Christian church which supported Aristotle's doctrines throughout the Middle Ages. /8

- Galileo was the first to combine empirical knowledge with mathematics and is therefore seen as the father of modern science. /8

- The birth of modern science was preceded and accompanied by a development of philosophical thought which led to an extreme formulation of the spirit/matter dualism. This formulation appeared in the seventeenth century in the philosophy of René Descartes who based his view of nature on a fundamental division into two separate and independent realms: that of mind (res cogitans), and that of matter (res extensa). The 'Cartesian' division allowed scientists to treat matter as dead and completely separate from themselves, and to see the material world as a multitude of different objects assembled into a huge machine. /8

- This inner fragmentation mirrors our view of the world 'outside,' which is seen as a multitude of separate objects

CHAPTER THREE

and events. The natural environment is treated as if it consisted of separate parts to be exploited by different interest groups. The fragmented view is further extended to society, which is split into different nations, races, religions and political groups. /8

- In contrast to the mechanistic Western view, the Eastern view of the world is 'organic.' For the Eastern mystic, all things and events perceived by the senses are interrelated, connected, and are but different aspects or manifestations of the same ultimate reality. /10

- In the Eastern view, then, the division of nature into separate objects is not fundamental and any such objects have a fluid and ever-changing character. The Eastern / world view is therefore intrinsically dynamic and contains time and change as essential features. The cosmos is seen as one inseparable reality—forever in motion, alive, organic; spiritual and material at the same time. /10-11

- Rational knowledge is derived from the experience we have with objects and events in our everyday environment. It belongs to the realm of the intellect, whose function is to discriminate, divide, compare, measure and categorize. In this way, a world of intellectual distinctions is created; of opposites which can exist only in relation to each other, which is why Buddhists call this type of knowledge 'relative.' /14

- For most of us it is very difficult to be constantly aware of the limitations and of the relativity of conceptual knowledge. Because our representation of reality is so much easier to grasp than reality itself, we tend to confuse the two and to take our concepts and symbols for reality. It is one of the main aims of Eastern mysticism to rid us of this confusion. Zen Buddhists say that a finger is needed to point to the moon, but that we should not trouble ourselves with the finger once the moon is recognized. /15

- In the West, the semanticist Alfred Korzybski made exactly the same point with his powerful slogan, 'The map is not the territory.' /16

- What Eastern mystics are concerned with is a direct experience of reality which transcends not only intellectual thinking, but also sensory perception. /16

- Eastern mysticism has developed several different ways of dealing with the paradoxical aspects of reality. Whereas they are bypassed in Hinduism through the use of mythical language, Buddhism and Taoism tend to emphasize the paradoxes rather than conceal them. /35

- Zen Buddhists have a particular knack for making a virtue out of the inconsistencies arising from verbal communication, and with the koan system they have developed a unique way of transmitting their teachings completely nonverbally. Koans are carefully devised non-sensical riddles which are meant to make the student of Zen realize the limitations of logic and reasoning in the most dramatic way. The irrational wording and paradoxical content of these riddles makes it impossible to solve them by thinking. They are designed precisely to stop the thought process and thus to make the student ready for the nonverbal experience of reality. /35

- One of the best koans, because the simplest, is Mu. This is its background: A monk came to Joshu, a renowned Zen master in China hundreds of years ago, and asked: 'Has a dog Buddha-nature or not?' Joshu retorted, 'Mu!' Literally, the expression means 'no' or 'not', but the significance of Joshu's answer does not lie in this. Mu is the expression of the living, functioning, dynamic Buddha-nature. What you must do is discover the spirit or essence of this Mu, not through intellectual analysis but by search into your innermost being. Then you must demonstrate before me, concretely, and vividly, that you understand Mu as living truth, without recourse to conceptions, theories, or abstract explanations. Remember, you can't understand Mu through ordinary cognition; you must grasp it directly with your whole being. (Referencing Kapleau, Three Pillars of Zen, Boston: Beacon Press, 1967, p. 135). /36

- Questions about the essential nature of things were answered in classical physics by the Newtonian mechanistic model of the universe which, much in the same way as the Democritean model in ancient Greece, reduced all phenomena to the motions and interactions of hard, indestructible atoms. The properties of these atoms were

CHAPTER THREE

abstracted from the macroscopic notion of billiard balls, and thus from sensory experience. Whether this notion could actually be applied to the world of atoms was not questioned. Indeed, it could not be investigated experimentally. /37

- The discoveries of modern physics necessitated profound changes of concepts like space, time, matter, object, cause and effect, etc.; and since these concepts are so basic to our way of experiencing the world, it is not surprising that the physicists who were forced to change them felt something of a shock. Out of these changes emerged a new and radically different world-view, still in the process of formation by current scientific research. /42

- The world-view which was changed by the discoveries of modern physics had been based on Newton's mechanical model of the universe. This model constituted the solid framework of classical physics. It was indeed a most formidable foundation supporting, like a mighty rock, all of science and providing a firm basis for natural philosophy for almost three centuries. The stage of the Newtonian universe, on which all physical phenomena took place, was the three-dimensional space of classical Euclidean geometry. It was an absolute space, always at rest and unchangeable. /43

- The mechanistic view of nature is ... closely related to a rigorous determinism. The giant cosmic machine was seen as being completely causal and determinate. All that happened had a definite cause and gave rise to a definite effect, and the future of any part of the system could—in principle—be predicted with absolute certainty if its state at any time was known in all details. (...) The philosophical basis of this rigorous determinism was the fundamental division between the I and the world introduced by Descartes. As a consequence of this division, it was believed that the world could be described objectively, i.e., without ever mentioning the human observer, and such an objective description of nature became the ideal of all science. /45

- Einstein strongly believed in nature's inherent harmony, and his deepest concern throughout his scientific life was to find a unified foundation of physics. He began to move toward his goal by constructing a common framework for electrodynamics and mechanics, the two separate theories of classical physics. This framework is known as the special theory of relativity. It unified and completed the structure of classical physics, but at the same time it involved drastic changes in the traditional concepts of space and time and undermined one of the foundations of the Newtonian world view. /50

- The whole development started when Max Planck discovered that the energy of heat radiation is not emitted continuously, but appears in the form of 'energy packets.' Einstein called these energy packets 'quanta' and recognized them as a fundamental aspect of nature. He was bold enough to postulate that light and every other form of electromagnetic radiation can appear not only as electromagnetic waves, but also in the form of these quanta. The light quanta, which gave quantum theory its name, have since been accepted as bona fide particles of a special kind, however, massless and always traveling with the speed of light. (...) At the subatomic level, matter does not exist with certainty at definite places, but rather shows 'tendencies to exist,' and atomic events do not occur with certainty at definite times and in definite ways, but rather show 'tendencies to occur.' In the formalism of quantum theory these tendencies are expressed as probabilities and are associated with mathematical quantities which take the form of waves. This is why particles can be waves at the same time. /56

- The basis of Krishna's spiritual instruction, as of all Hinduism, is the idea that the multitude of things and events around us are but different manifestations of the same ultimate reality. This reality, called Brahman, is the unifying concept which gives Hinduism its essentially monistic character in spite of the worship of numerous gods and goddesses. /77

- Maya ... does not mean that the world is an illusion, as it is often wrongly stated. The illusion merely lies in our point of view, if we think that the shapes and structures, things and events, around us are realities of nature, instead of realizing that they are concepts of our measuring and categorizing minds. Maya is the illusion of taking these concepts for reality, of confusing the map with the territory. /78

- Recognizing the relativity of good and bad, and thus of all moral standards, the Taoist sage does not strive for the good

CHAPTER THREE

but rather tries to maintain a dynamic balance between good and bad. /103

- When Po-chang was asked to define Zen, he said 'When hungry, eat, when tired, sleep.' Although this sounds simple and obvious, like so much in Zen, it is in fact quite a difficult task. To regain the naturalness of our original nature requires long training and constitutes a great spiritual achievement. /110

- The notion that all opposites are polar—that light and dark, winning and losing, good and evil, are merely different aspects of the same phenomenon—is one of the basic principles of the Eastern way of life. Since all opposites are interdependent, their conflict can never result in the total victory of one side, but will always be a manifestation of the interplay between the two sides. In the East, a virtuous person is therefore not one who undertakes the impossible task of striving for the good and eliminating the bad, but rather one who is able to maintain a dynamic balance between good and bad. /131

- Western society has traditionally favored the male side rather than the female. Instead of recognizing that the personality of each man and of each woman is the result of an interplay between male and female elements, it has established a static order where all men are supposed to be masculine and all women feminine, and it has given men the leading roles and most of society's privileges. This attitude has resulted in an over-emphasis of all the yang—or male—aspects of human nature: activity, rational thinking, competition, aggressiveness, and so on. The yin—or female— modes of consciousness, which can be described by words like intuitive, religious, mystical, occult, or psychic, have constantly been suppressed in our male-oriented society. /133

- Contraria sunt complementa (Opposites are complementary), Niels Bohr acknowledged the profound harmony between ancient Eastern wisdom and modern Western science. /146

- Our notions of space and time figure prominently on our map of reality. They serve to order things and events in our environment and are therefore of paramount importance not only in our everyday life, but also in our attempts to understand nature through science and philosophy. There is no law of physics which does not require the concepts of space and time for its formulation. The profound modification of these basic concepts brought about by relativity theory was therefore one of the greatest revolutions in the history of science. /147

> Greek natural philosophy was, on the whole, essentially static and largely based on geometrical considerations. It was, one could say, extremely 'non-relativistic,' and its strong influence on Western thought may well be one of the reasons why we have such great conceptual difficulties with relativistic models in modern physics. The Eastern philosophies, on the other hand, are 'space-time' philosophies, and thus their intuition often comes very close to the views of nature implied by our modern relativistic theories. /159

Chapter Four
Green Politics

Green Politics

The Global Promise
How the Greens are transforming the political culture of Europe and inspiring a worldwide movement that can change the course of America's future
With Charlene Spretnak
Santa Fe, NM: Bear & Company, 1986

Green Politics shows Fritjof Capra as a motivated, intuitive and highly knowledgeable political thinker. I do not think that all Capra fans are interested in this book for it's not always a captivating read, given the little critter of political programs, but for me as a biographical researcher, this book next to 'Uncommon Wisdom' reveals very much about the scientist Fritjof Capra, when and how he is *more* than a scientist!

As I am myself a rather a-political person, I bought this book only for including it here in this biographical book about Fritjof Capra. I had not even known about it, and it was through a correspondence with the *Center of Ecoliteracy* in Berkeley, California that I was given the reference to this book and the one I am going to review further down in Chapter 5, 'Belonging to the Universe.'

I am grateful to have received that advice for if I had not included these books here, the picture the reader gets about Fritjof Capra would be incomplete, to say the least!

Once again, the openness and truthful language of Capra convinces when it goes to revealing some really ugly details about modern society, the militarization of culture, and the role the military-industrial complex plays in funding science, and in *not* funding the science that does not serve the interests of war and conflict between nations.

I do not know any other scientist who is as outspoken about the unilaterality of political purpose, and its often veiled target to uncompromisingly mislead voters if not the entire media world about its true motives! Perhaps, like me, you won't be a-political anymore after reading this book?!

Honestly, despite the fact that I am trained as an international lawyer and have studied international politics as well, I always believed that politics is that 'dirty business' I better not care about. Wrong!

If people like you and me do not care about politics, who will? Should we leave it to the uneducated masses to vote our politicians? And even prior to voting for certain politicians, should we not first of all vote political agendas? And the 'green agenda' really is interesting, because it is different, *more* different than all the others are from one another. And it's daring in many ways, for without changing some base paradigms, this agenda will probably never see its day in parliament …

Now, as I proceeded in my previous review, I will present to you here quotes that allow you to make up your mind about the book, without me going to bore you with paraphrasing over and over again.

Chapter 2

Principles of a New Politics

The Greens begin their Federal Program by explaining why a new politics is necessary:

The Establishment parties in Bonn behave as if an infinite increase in industrial production were possible on the finite planet Earth. According to their own statements, they are leading us to a hopeless choice between the nuclear state or nuclear war, between Harrisburg or Hiroshima. The worldwide ecological crisis worsens from day to day: natural resources become more scarce; chemical waste dumps are subjects of scandal after scandal; whole species of animals are exterminated; entire varieties of plants become extinct; rivers and oceans change slowly into sewers; and humans verge on spiritual and intellectual decay in the midst of a mature, industrial, consumer society. It is a dismal inheritance we are imposing on future generations.

We represent a total concept, as opposed to the one-dimensional, still-more-production brand of politics. Our policies are guided by long-term visions for the future and are grounded on four basic principles: ecology, social responsibility, grassroots democracy, and nonviolence. /29-30

Green politics, then, is inherently holistic in theory and practice. It is based on ecological, or 'network,' thinking, a term used frequently by the Greens. Ecological thinking also includes the realization that the seemingly rigid structures we perceive in our environment are actually manifestations of underlying processes, of nature's continual dynamic flux. Interrelatedness and ongoing process are the lessons the Greens take from and apply to the ecosystems surrounding us. They support 'soft' energy production (such as solar power) that works with the cycles of the sun, the water, and the wind, and the flow of the rivers. They call for the development of appropriate technology that reflects our interdependence with the Earth. They advocate regenerative agriculture that replenishes the soil and incorporates natural means of pest control. Above all, the Greens demand a halt to our ravaging of natural 'resources' and our poisoning of the biosphere through the dumping of toxic wastes, the accumulation of so-called acceptable levels of radiation exposure, and the pollution of the air. /30-31

Although Western culture has been dominated for several hundred years by a conceptualization of our bodies, the body politics, and the natural world as hierarchically arranged aggregates of discrete components, that world view is giving way to the systems view, which is supported by the most advanced discoveries of modern science and which is deeply ecological. In its early stages, during the 1940s, systems theory was closely linked with the study of control and regulatory mechanisms of complex machines and electronic systems. During the past decade, however, the focus has shifted to the study of living systems: living organisms, social systems, and ecosystems. The emergent systems view of life was developed by a number of scientists from various disciplines: Ilya Prigogine, Erich Jantsch, Gregory Bateson, Humberto Maturana, and Manfred Eigen, to name but a few. /31

The systems view involves looking at the world in terms of relationships and integration. Systems are integrated wholes whose properties cannot be reduced to those of smaller units. Whereas for two thousand years most of Western science has concentrated on reducing the world to its basic building blocks, the systems approach emphasizes principles of organization. Examples of living systems abound in nature. Every organism—from the smallest bacterium through the wide range of plants and animals to humans—is an integrated whole and thus a living system. Cells are living systems, and so are the various tissues and organs of the body. The same characteristics of wholeness are exhibited by social systems—such as a family or a community—and by ecosystems that consist of a variety of organisms and inanimate matter in mutual interaction. /31

Chapter 9

Possibilities for Green Politics in America: 1983

The roots of Green ideas in American culture reach back to our earliest origins. For more than 20,000 years Native Americans have maintained a deeply ecological sense of the subtle forces that link humans and nature, always emphasizing the need for balance and for reverence toward Mother Earth. Spiritual values are inherent in their politics, as they were for the many colonists who came to this land for the protection of religious pluralism. The Founding Fathers of our government, who were familiar with the federal system of the Iroquois nation, created a democratic federalism that reflects the shared values comprising national identity but entrusts extensive powers to the states and to the people's representatives, who can block the designs of federal authoritarianism. The young nation spawned a network of largely self-sufficient communities that flourished through individual effort and cooperation—the barn raisings, the quilting bees, the town meetings. Yet local self-sufficiency and self-determination eventually gave way to control by such huge institutions as the federal bureaucracy, the military establishment, massive corporations, big labor unions, the medical establishment, the education system, institutionalized religion, and centralized technology. / 193-194

The ecology and peace movements have discovered their common ground, the feminists have held ecofeminist conferences and peace actions, and countless networks working toward comprehensive, nonviolent social change have developed. Most of these people are working with a 'big picture' orientation, rather than single-focus problem solving. They are among the fifteen million adult Americans who, according to recent studies by the research institute SRI International, are basing their lives fully or partially on such values as frugality, human scale, self-determination, ecological awareness, and personal growth. In addition, the holistic health movement seriously challenges the mechanistic approach of the medical establishment. Many churches are now reinterpreting the Scriptural charge to 'have dominion over the earth,' reading it as a call to stewardship rather than exploitation, and some are even going beyond stewardship to deep ecology. Numerous positive steps have been taken toward realizing that our existence is part of a subtle web of interrelationships—yet these fall far short of creating an effective political manifestation of the new paradigm. / 195

We believe it is essential that Green ideas enter the American political debate at all levels. Currently the Democratic and Republican parties struggle fruitlessly to apply outdated and irrelevant concepts and priorities to our burgeoning crisis. They are unable to respond effectively to changing conditions such as the end of the fossil-fuel age and the growth of global interdependence and so are leading us toward disaster. As the quality of life in this country declines and hardships in the Third World increase, the old-paradigm parties are losing credibility. Ronald Reagan was elected president with only 28 percent of the eligible vote; hopelessness and fearful apathy carried the majority. Behind the rhetoric of both parties, it is apparent that one of their shared functions is to remain nonideological, to diffuse dissent rather than standing for a coherent program. / 195-196

To consider the possibilities for Green politics in the United States, we should first reflect on the lessons from West Germany—with the understanding that Green politics here, as in other countries, must grow from our own cultural and political tradition, and from our current situation. / 196

Once they [the Greens] won seats in the legislative bodies, a great deal of their attention shifted from evolving responses and comprehensive positions to internal power struggles and ongoing debates on legislative strategy. / 196

Chapter 10
Green Politics in the United States: 1986

Ten Key Values

1. Ecological Wisdom

How can we operate human societies with the understanding that we are *part* of nature, not on top of it? How can we live within the ecological and resource limits of the planet, applying our technological knowledge to the challenge of an energy-efficient economy? How can we build a better relationship between cities and countryside? How can we guarantee rights of nonhuman species? How can we promote sustainable agriculture and respect for self-regulating natural systems? How can we further biocentric wisdom in all spheres of life? / 230

2. Grassroots Democracy

How can we develop systems that allow and encourage us to control the decisions that affect our lives? How can we ensure that representatives will be fully accountable to the people who elected them? How can we develop planning mechanisms that would allow citizens to develop and implement their own preferences for policies and spending priorities? How can we encourage and assist the 'mediating institutions'—family, neighborhood organization, church group, voluntary association, ethnic club—recover some of the functions now performed by government? How can we relearn the best insights from American traditions of civic vitality, voluntary action, and community responsibility? /230

3. Personal and Social Responsibility

How can we respond to human suffering in ways that promote dignity? How can we encourage people to commit themselves to lifestyles that promote their own health? How can we have a community controlled education system that effectively teaches our children academic skills, ecological wisdom, social responsibility, and personal growth? How can we resolve interpersonal and intergroup conflicts without just turning them over to lawyers and judges? How can we take responsibility for reducing the crime rate in our neighborhoods? How can we encourage such values as simplicity and moderation? /230-231

4. Nonviolence

How can we, as a society, develop effective alternatives to our current patterns of violence, at all levels, from the family and the street to nations and the world? How can we eliminate nuclear weapons from the face of the Earth without being knowledgeable about the intentions of other governments? How can we most constructively use nonviolent methods to oppose practices and policies with which we disagree and in the process reduce the atmosphere of polarization and selfishness that is itself a source of violence? / 231

5. Decentralization

How can we restore power and responsibility to individuals, institutions, communities and regions? How can we encourage the flourishing of regionally based culture rather than a dominant monoculture? How can we have a decentralized, democratic society with our political, economic, and social institutions locating power on the smallest scale (closest to home) that is efficient and practical? How can we redesign our institutions so that fewer decisions and less regulation over money are granted as one moves from the community toward the national level? How can we reconcile the need for community and regional self-determination with the need for appropriate centralized regulation in certain matters? / 231

6. Community-based Economics

How can we redesign our work structures to encourage employee ownership and workplace democracy? How can we develop new economic activities and institutions that will allow us to use our new technologies in ways that are humane, freeing, ecological, and accountable and responsive to communities? How can we establish some form of basic economic security, open to all? How can we move beyond the narrow 'job ethic' to new definitions of 'work,' jobs,' and 'income' that reflect the changing economy? How can we restructure our patterns of income distribution to reflect the wealth created by those outside the formal, monetary economy: those who take responsibility for parenting, housekeeping, home gardens, community volunteer work, etc.? How can we restrict the size and concentrated power of corporations without discouraging superior efficiency or technological innovation? / 231-232

7. Postpatriarchal Values

How can we replace the cultural ethics of dominance and control with more cooperative ways of interacting? How can we encourage people to care about persons outside of their own group? How can we promote the building of respectful, positive, and responsible relationships across the lines of gender and other divisions? How can we encourage a rich, diverse political culture that respects feelings as well as relational approaches? How can we proceed with as much respect for the means as the end (the process as much as the products of our efforts)? How can we learn to respect the contemplative, inner part of life as much as the outer activities? / 232

8. Respect for Diversity

How can we honor cultural, ethnic, racial, sexual, religious, and spiritual diversity within the context of individual responsibility to all beings? While honoring diversity, how can we reclaim our country's finest shared ideals: the dignity of the individual, democratic participation, and liberty and justice for all? / 232

9. Global Responsibility

How can we be of genuine assistance to grassroots groups in the Third World? What can we learn from such groups? How can we help other countries make the transition to self-sufficiency in food and other basic necessities? How can we cut our defense budget while maintaining an adequate defense? How can we promote these ten Green values in the reshaping of global order? How can we reshape world order without creating just another enormous nation-state? / 232-233

10. Future Focus

How can we induce people and institutions to think in terms of the long-range future, and not just in terms of their short-range selfish interest? How can we encourage people to develop their own visions of the future and move more effectively toward them? How can we judge whether new technologies are socially useful—and use those judgments to shape our society? How can we induce our government and other institutions to practice fiscal responsibility? How can we make the quality of life, rather than open-ended economic growth, the focus of future thinking? / 233

Chapter Five

Belonging to the Universe

Belonging to the Universe

Explorations at the Frontiers of Science & Spirituality
Fritjof Capra & David Steindl-Rast
With Thomas Matus
San Francisco: Harper & Row, 1991

This highly interesting book is difficult to review for it's a recorded conversation between two theologians and a scientist. Once again, this event and the resulting book reflects Capra's open-mindedness and his interest in exploring areas where he is not an expert, but is eager to engage those who know more in dialogue. As for this book, the result of the idea is once again absolutely convincing.

While this book is not an easy read for the discussion is highly philosophical, yet at times technical and concerned with elucidating terminology, it does not for that matter require any special knowledge of theology.

What astonished me in particular was the high level of *synergy and human resonance* between the three men, which is certainly extraordinary as it doesn't happen on a daily basis that scientists discuss with theologians the frontiers of both science and theology.

The book begins with a Preview, an excellent idea, even before the Introduction. You got two columns here over 5 pages that reflect paradigm-changing details from science and theology. Let me quote the sub-headers here. The general header is: *New-Paradigm Thinking in Science by Fritjof Capra* versus *New Paradigm Thinking in Theology, a paraphrase by Thomas Matus and David Steindl-Rast.*

1. Shift from the Part to the Whole vs. Shift from God as Revealer of Truth to Reality as God's Self-Revelation;

2. Shift from Structure to Process vs. Shift from Revelation as Timeless Truth to Revelation as Historical Manifestation;

3. Shift from Objective Science to 'Epistemic Science' vs. Shift from Theology as an Objective Science to Theology as a Process of Knowing;

4. Shift from Building to Network as Metaphor of Knowledge vs. Shift from Building to Network as Metaphor of Knowledge;

5. Shift from Truth to Approximate Descriptions vs. Shift in Focus from Theological Statements to Divine Mysteries.

The book is a highly collaborative effort. Even, and this is unusual, the Introduction is not written by an editor, but is already part of the discussion and proceeds in dialogue.

As I consider it very difficult to paraphrase anything of the discussion, I have chosen to provide the reader with two quite interesting points as samples of the nature of the entire discussion. I think they are quite illustrative for that purpose.

3 Paradigms in Science and Theology (pp. 33-39)
Paradigms in science and society

Fritjof: We have talked about the purposes and the methods of science and theology. Now I would like to introduce a historical perspective and talk about how scientific theories develop and how knowledge is accumulated in science. As you know, until recently the belief was that there is a steady accumulation of knowledge; that theories gradually get more and more comprehensive and more and more accurate.

Thomas Kuhn introduced the idea of paradigms and paradigm shifts, which says there are these periods of steady accumulation, which he calls normal science, but then there are periods of scientific revolutions where the paradigm changes. A scientific paradigm, according to Kuhn, is a constellation of achievements—by that he means concepts, values, techniques, and so on—shared by a scientific community and used by that community to define legitimate problems and solutions.

So this means that behind the scientific theory is a certain framework within which science is pursued. And it's important to notice that this framework includes not just concepts but also values and techniques. So the activity of doing science is part of the paradigm. The attitude of domination and control, for example, is part of a scientific paradigm. /33-34

David: Would you say it is part of the paradigm? Or is it a force that conditions the paradigm? /34

Fritjof: It is part of the paradigm, because it's part of the values underlying the scientific theories. The values are part of the paradigm. So a paradigm to Kuhn and to me is more than a worldview, more than a conceptual framework, because it includes values and activities. To make this clearer, let me show you how I enlarged that, following Marilyn Ferguson and Willis Harman and other people who have often used 'paradigm' in a larger sense. I have taken the Kuhnian definition and enlarged it to that of a social paradigm.

A social paradigm, for me, is a constellation of concepts, values, perceptions, and practices, shared by a community that forms a particular vision of reality that is the basis of the way the community organizes itself. It's necessary for a paradigm to be shared by a community. A single person can have a worldview, but a paradigm is shared by a community. /34

David: And why do you speak only about communal organizations and not about the whole life of the community? Why do you focus on the organization only? Why not on values? /34

Fritjof: I have not explored the difference between paradigm and culture. You could say the basis of the whole life is the culture. The two are closely related, but I have not gone into this. /35

Now Kuhn, of course, uses the term in a narrower sense, and within science he talks about different paradigms. I use it in a very broad sense, the sort of overarching paradigm underlying the organization of a certain society or the organization of science in a certain scientific community. /35

David: I asked about values because I thought you were still talking about a paradigm shift within a particular science. There the values would of course be implicit, not at all explicit. /35

Fritjof: The entire notion of the paradigm is implicit in the periods of normal science, and it's very difficult to delineate the paradigm and to show where the limitations are, where its borders are. It's only in times when the paradigm changes that you see its limitations, and, in fact, it changes *because* of these limitations. Kuhn has written very extensively about that. When there are problems, which he calls anomalies, that can no longer be solved within the dominant paradigm, these shifts occur. And of course it takes a while until these problems actually force people to shift.

In physics, for example, the most recent paradigm shift began in the 1920s when various problems connected with atomic structure could not be solved in terms of Newtonian science. And what I am saying in my book *The Turning Point* is that now we are in a situation in society where the social paradigm has reached its limitations. These limitations are the threat of nuclear war, the devastation of our natural environment, the persistence of poverty around the world—all these very severe problems that can no longer be solved in the old paradigm.

Kuhn, by the way, speaks of a pre-paradigmatic period where there are competing views. One of them will then become the dominant paradigm, shared by the scientific community. In society or, say, in the human family, this is different, because we do have different coexisting social paradigms. The Islamic social paradigm is different from the Japanese or from the American. So the same group of phenomena—like economics, politics, and social life—will be understood in terms of different coexisting paradigms. /35

David: Can you explain why different paradigms can coexist in a social context and not in science? /35

Fritjof: There *could* be different coexisting paradigms also in science, and there were in the past, but not since the rise of European science in the seventeenth century. Wherever people do science now, in the modern sense of the term, they would do science according to the European paradigm, whether it's in Japan or China or Africa. Many scientists say they have been brainwashed to do that. They could do science within another paradigm, but they don't. There is a certain colonization of scientists by European and American science. Now it's America, but the roots, of course, are in European science. Whereas in social matters, there's not only so much dominance of a single paradigm. Different cultures coexist. In science we do not find different cultures coexisting; there's basically one scientific culture. /36

David: What you said just now is really very important, yet it often goes unnoticed that even in science it would be possible to have different paradigms next to one another. It is almost accidental that there is one scientific paradigm, due to the colonialism of Western science. It need not be so. You said that scientists could do science in a different paradigm. This is important. However, people often say: 'Well this is just the strength of science, that is unifies. In science there can be no contradictions. Science is the rock-bottom basis for all truth,' and so on. /36

Fritjof: But you see, science is pursued within the larger paradigm. So, for instance, if two scientific groups worked on the Strategic Defense Initiative (SDI) project, they would get very similar results. They would construct laser beams for use in outer space, space stations, killer satellites, and so on. Although the results would differ somewhat, as they do in science when it is done in different countries, more or less the same conclusions would be reached. But you could easily imagine that in one culture it would be absolutely out of the question even to work on such a project, because the values would be different. /36

David: That's what I want to emphasize, the connection between the social and the scientific paradigm: what kind of society we live in determines what kind of science we are going to have. /36

Fritjof: Yes, the scientific paradigm is embedded in the social paradigm. /36

David: Much more so than people realize. Now, let me ask you something else. I have long been fascinated with the concept of ether. Ether played such an important role in the history of science up to the late nineteenth century at least. Now it has been dropped completely. What happened? Why was it needed, and why is it not needed anymore? Maybe we can find a parallel here with certain theological concepts that also seemed urgently needed at one time and are now no longer necessary. That seems to be a typical phenomenon in times of paradigm changes. /37

Fritjof: Yes, it is. This phenomenon of concepts that are needed during a certain time and then are not needed anymore happens again and again in science. We guild models and then we discard them, because we have better models. Then finally we have a complete theory that is not discarded. It will be superseded by better theories but will still be valid within its range of applicability.

Among the scientific concepts that were discarded then a new model was adopted, the ether is perhaps the most famous, and rightly so, because the shift of perception that allowed us to discard the concept of an ether marks the beginning of twentieth-century physics.

This is a fascinating subject. It begins with the question of the nature of light, and it is a very powerful illustration of the fact that such nature of light, and it is a very powerful illustration of the fact that such an everyday experience as sunlight reaching the Earth is something that goes beyond our powers of imagination. We have no way to imagine how sunlight reaches the Earth. Although people normally are unaware of this, this question bot scientists into modern physics.

In the nineteenth century, Michael Faraday and Clerk Maxwell developed a comprehensive theory of electromagnetism, which culminated in the discovery that light consists of rapidly alternating electric and magnetic fields that travel through space as waves. These fields are nonmechanical entities, and Maxwell's equations, which describe their exact behavior, were the first theory that went beyond Newtonian mechanics. That was the great triumph of nineteenth-century physics.

However, when Maxwell made his discovery, he was immediately faced with a problem. If light consists of electromagnetic waves, how can these waves travel through empty space? We know from our experience and from the theory of waves that every wave needs a medium. A water wave needs the water that is disturbed and then moves up and down as the wave passes through. A sound wave needs the particles of air, vibrating as the wave passes through. Without air or some other material substance, there is no sound. But light waves travel through empty space, where there is no medium to transmit the vibrations. So what is vibrating in a light wave?

This is what led scientists to invent the ether. They said, 'There's no air, but there is an invisible medium, called ether, in which light waves travel.' This ether had to have fancy properties. For example, it had to be a weightless and perfectly elastic substance. You see, when water waves travel, they diminish because of friction, but light waves don't. So the ether had to be perfectly elastic without any friction. Scientists at the beginning of the twentieth century could not bring themselves to abandon the notion of an ether, in spite of its strange properties, because this mechanistic image of a wave needing a medium was so firmly ingrained in their minds.

It took an Einstein to say that there was no ether, that light is a physical phenomenon in its own right, which doesn't need a medium. It doesn't need a medium, said Einstein, because it manifests not only as waves but also as particles, which can travel through empty space. He called those particles of light *quanta*, which gave the name to quantum theory, the theory of atomic phenomena.

The struggle with the question, in what sense exactly is a quantum of light a particle and in what sense is it a wave? is the story of quantum theory, spanning the first three decades of the century. At the end of that exciting period, physicists understood that light waves are really 'probability waves'—that is, abstract mathematical patterns that give you the probability of finding a particle of light (which today we call *photon*) in a particular place when you look for it. These probability patterns are wave patterns that travel through empty space. So, without going into further details, the end of the story is that light is both particles and waves, and the ether is no longer needed. /37-38

David: So in physics we once had a concept that seemed absolutely indispensable, and then it dropped away. I think there are parallels to this phenomenon in theology. /38

Thomas: The classic example of an unnecessary doctrine within common Christian theological thought is the geocentric universe. In order to uphold the truthfulness of the Bible, medieval theologians thought it necessary to posit an immobile Earth at the center of a moving cosmos. During the Renaissance, Copernicus and others elaborated another theory: that the Earth is not the center but is moving around the sun. Galileo sustained the Copernican thesis. At the same time, however, Galileo was an ardent Catholic who desired to remain in full communion with the Christian Church. He was not unsophisticated in theology. He had read the Bible and felt the need to explain the relationship between science and theology, or better yet, between scientific language and biblical language. /38-39

Fritjof: What was the theological problem? /39

Thomas: Theologians believed that since the Bible said, 'The sun stood still,' for instance, it was necessary, in order not to cast doubt on the truth of Holy Scripture, to assume that the sun moved around the Earth. /39

David: One mistook poetic language for scientific reporting. /39

Thomas: Galileo said that this verse of the Bible, 'The sun stood still,' was a religious statement. The language it uses is the language of the common people; it addresses the masses, while science is for people who speak a different, more sophisticated language, the language of mathematics. The purpose of science is not to fulfill people's religious needs but to gain knowledge about the universe and to build the great edifice of empirical knowledge. This statement, with greater refinement, is one that any biblical scholar would make today. /39

Fritjof: So what was the concept that was no longer needed? /39

Thomas: The concept that was no longer needed was that of the immobile Earth. Ultimately theologians came to the conclusion that the Bible was not a scientific textbook, a source of answers to our questions about the physical universe. /39

Fritjof: Could one say that the Bible speaks in terms of metaphors and models as we do in science? The metaphors of the Bible point toward religious truth, but they are not the full truth. So the metaphor should not be confused with the truth toward which the metaphor points. /39

New thinking and new values (pp. 73-77)

Fritjof: I also would like to show you a striking and somewhat surprising pattern of the paradigm change, a connection between thinking and values. It turns out that the old thinking and the old values hang together, are very closely intertwined. And correspondingly the new thinking and the new values are closely intertwined.

In both cases, thinking and values, there is a shift of emphasis from self-assertion to integration. That's the best way I've found to characterize those groups of modes of thinking and of values.

In thinking, the shift has been from the rational to the intuitive. Rational thinking consists in compartmentalizing, distinguishing, categorizing. That's very much connected with the whole notion of the self as a distinct category, so it's clearly self-assertive. Analysis is this method of distinguishing and categorizing, and there has been a shift from analysis to synthesis; a shift from reductionism to holism, from linear thinking to nonlinear thinking.

As far as values are concerned, you have a shift from competition to cooperation—very clearly a shift from self-assertion to integration; from expansion to conservation; from quantity to quality; from domination to partnership (as Riane Eisler has emphasized).

Now, if you look at this from the systems point of view, from the point of view of living systems, you realize that since all living systems are embedded in larger systems, they have this dual nature that Arthur Koestler called a Janus nature. On the one hand, a living system is an integrated whole with its own individuality, and it has a tendency to assert itself and to preserve its individuality. As part of the larger whole, it needs to integrate itself into that larger whole. It's very important to realize that those are opposite and contradictory tendencies. We need a dynamic balance between them, and that's essential for physical and mental health. The Chinese picked this up with great intuitive power. In order to have a healthy life, you need to assert yourself and you need to integrate yourself.

I think culturally and socially you can say that the pendulum has swung between those two tendencies. For instance, the Middle-Ages were characterized by a lot of integration but also by a lack of self-assertion. /73-74

David: Overemphasis on integration. /74

Fritjof: But then with the Renaissance, you have the emergence of individuality. Then it went further in the nineteenth century, and later, especially here in America, you have an overemphasis on individuality—the cowboy ethic, rugged individualism, and so on.

The emergence of individuality gave rise to individualism all over the Western world, but you had socialism as a counter tendency. This then went too far in the socialist countries, which are now looking for a balance. Humanism, of course, is the key word for the emergence of individuality. And so Gorbachev and several Marxist philosophers before him have been talking about 'new humanism.' In Prague in 1968, Dubcek introduced a 'socialism with a human face.' Similarly, E.F. Schumacher was talking about technology with a human face, because technology had become so oppressive.

I have taken this interplay between these tendencies, self-assertion and integration, as my framework to talk about values in contemporary society, where you can see consistently an overemphasis of self-assertion and neglect of integration.

The other important connection is to the patriarchal value system, because the self-assertive values and modes of thinking are the masculine ones. Whether this is biological or cultural is a very tricky question, and I don't want to go into this. But in most cultures, and certainly in our culture, the self-assertive ways of thinking and the self-assertive values have been associated with men, with manliness, and have been given political power. /74-75

Thomas: Would you say that, as ways of knowing, the theories associated with self-assertion give different results from those associated with integration? In other words, you arrive at a different content of knowledge, depending upon which mode of thinking you use. If you use the rational-analytic-reductionist-linear mode, you're going to learn certain things about nature but not others. Whereas if you use the intuitive-synthetic-holistic-nonlinear mode, you'll learn other things. /75

Fritjof: Yes, but you also have to realize that you cannot use just one. In science you always need both.

David: Is there not another term that one could use rather than rational to indicate the polar opposite of intuitive? /75

Thomas: I think the nearest thing would be conceptual and non conceptual kinds of knowledge. There's also an intuitive conceptualization, but concepts are most often formed through rational process, as the fruit of deductive reasoning. /75

David: I'm very sensitive to a danger implicit in expressing it in this way; namely, that you equate intuitive with irrational, and that would be terribly wrong. /76

Fritjof: Let me tell you what I mean without using any of these terms, and we'll come up with something. The self-assertive mode is a way of thinking that categorizes, that divides, that takes to pieces, that delineates. The other one is a way of perceiving nonlinear patterns, a synthesis of a nonlinear pattern. Intuition, to me, is an immediate perception of the whole, of a gestalt. /76

David: The very word *intuition* means that you 'look into' it. You take so deep a look that you see an inner coherence. /76

Fritjof: No, I wouldn't call it rational, because I can't talk about it. To me rational is what you can talk about. /76

Thomas: Then maybe you should call it not rational but discursive. /76

David: ... *discursive* and intuitive, that's a fine pair of opposite terms! Now I'm satisfied. Let's ask the question of ourselves: Is there a general shift in thinking and in values from self-assertion to integration also in theology? My intuitive response is Yes! Empathically so. Let's see if some analysis will prove this intuition correct. /76

Thomas: I think that, from a number of different viewpoints, this could be borne out in the contemporary theological discussion. For one thing, the apologetical and polemical thrust of most Positive-Scholastic theology tends to suggest the self-assertive mode. Whereas the ecumenical orientation of most contemporary or new-paradigm theology suggests the integrative. In other words, true fidelity to one's tradition requires a full and open understanding of other traditions. /76

David: Also, more specifically, there is this switch from theological propositions to story telling. Originally all theological insights were stories before they became propositions. Why don't we turn them into stories again? Many ask this question today. That means a switch from the discursive to the intuitive, the story is intuitive; from the analytic to the synthetic; the story is synthetic; from the reductive to the holistic, because the story is a whole, greater than the sum total of its parts. /76-77

Thomas: Of course, you wouldn't want to limit it to the narrative literary genre. You could also say that there is a shift from the propositional to the poetic or metaphorical.

David: Yes, or from the abstract to the experiential. All this fits in.

Fritjof: Story telling, by the way, was the preferred mode of Gregory Bateson, who was one of the key figures in the development of systems thinking. In his presentation Bateson was essentially a storyteller. His way of showing the connectedness of various patterns was through a story. /77

Chapter Six
The Turning Point

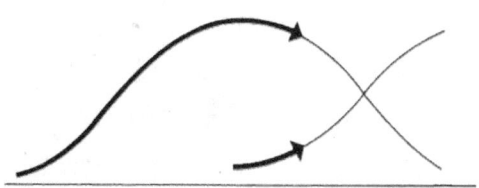

The Turning Point

Science, Society and the Rising Culture
New York: Simon & Schuster (Flamingo), 1987
Original author copyright 1982

Review

In my personal view, and contrary to what most critics say, it is lesser the *Tao of Physics* that is Capra's real strike of genius, but the present book because of the extrapolation of the holistic science concept developed in the Tao upon the whole value system of postmodern international culture, thereby suggesting our culture adopting and developing an entire new set of values.

Only a thinker who is both logically very precise, very knowledgeable about science history, and who has a metarational and integrated perception of life and the universe could do such a giant work.

The following quote shows the general direction that Capra took from the time of writing this book, and that will be especially present in his two subsequent books, *The Web of Life* and *The Hidden Connections*. It has been called *the systems view*; it simply is a sound holistic science paradigm that can be *practically applied to all scientific research,* and that promises to bring about scientific, social and later political results that are in accordance with human dignity, fostering the expansion of human consciousness and evolution. These solutions will be different from those we had in the past because they will be integrated and sustainable, and this both in the fields of science and culture:

> These problems (…) are systemic problems, which means that they are closely interconnected and interdependent. They cannot be understood within the fragmented methodology characteristic to our academic disciplines and government agencies. Such an approach will never solve any of our difficulties but will merely shift them around in the complex web of social and ecological relations. A resolution can be found only if the structure of the web is changed, and this will involve profound transformations of our social institutions, values, and ideas./6

One of the points that show Capra a genius is his mental flexibility. Contrary to many other scientists from the so-called *exact* scientific disciplines, he has an extraordinarily synthetic thinking ability which makes him sense paradigm shifts and developments in society long before they actually happen. Then, following his intuition, he puts his sharp rational mind in the forefront for collecting and arranging the information he needs to elucidate and deploy.

This is in full accordance with Einstein's saying that a problem can never be solved on the level of thought that brought it about. In fact, it's only through creative thinking and intuition that we find new solutions to our old problems, because we then relocate the thinker to a higher level of perspective.

This can be seen in the way Capra puts spotlights on trends and philosophical movements of old, to show the potential they had for forging the reigning worldview, or else for shifting that view and preparing the ground for a *paradigm shift*. In his book *Uncommon Wisdom*, the author explains what a paradigm is:

> A paradigm, for me, would mean the totality of thoughts, perceptions, and values that form a particular vision of reality, a vision that is the basis of a way of society organizes itself. /22

For example, Heraclites was one of those enlightened minds who showed us the volatile path of integrated wisdom, but he was not followed. Instead our science was to slavishly follow Aristotle, and in the East, the same happened when *Lao-tzu* was shunned by Chinese thinkers for giving the preference to the pedantic, moralistic and hair-splitting Confucius. This shows with some evidence that history is not linear, and that it is very much a function of the reigning paradigm.

In this revolutionary book, which to me was way more convincing than *The Tao of Physics,* Fritjof Capra carefully examines the reigning paradigms not only in science but more importantly so in the social sciences and evaluates them from a point of view of effectiveness and sustainability. He then works out a pathway in each and every case, and for each and every discipline, be it psychology or cancer research, how the particular discipline would be transformed by a applying a *systemliterate* perspective. He writes:

Studies of periods of cultural transformation in various societies have shown that these transformations are typically preceded by a variety of social indicators, many of them identical to the symptoms of our current crisis. They include a sense of alienation and an increase in mental illness, violent crime, and social disruption, as well as an increased interest in religious cultism—all of which have been observed in our society during the past decade. In times of historic cultural change these indicators have tended to appear one to three decades before the central transformation, rising in frequency and intensity as the transformation is approaching, and falling again after it has occurred. /7

The best summary of the entire huge tenor and conceptual novelty of the book can be found in these enlightening quotes:

> In truth, the understanding of ecosystems is hindered by the very nature of the rational mind. Rational thinking is linear, whereas ecological awareness arises from an intuition of nonlinear systems. One of the most difficult things for people in our culture to understand is the fact that if you do something that is good, then more of the same will not necessarily be better. This, to me, is the essence of ecological thinking. (…) Ecological awareness, then, will arise only when we combine our rational knowledge with an intuition for the nonlinear nature of our environment. Such intuitive wisdom is characteristic of traditional, nonliterate cultures, especially of American Indian cultures, in which life was organized around a highly refined awareness of the environment. (…) Biological evolution of the human species stopped some fifty thousand years ago. From then on,

evolution proceeded no longer genetically but socially and culturally, while the human body and brain remained essentially the same in structure and size. /25

Our progress, then, has been largely a rational and intellectual affair, and this one-sided evolution has now reached a highly alarming stage, a situation so paradoxical that it borders insanity. We can control the soft landings of space craft on distant planets, but we are unable to control the polluting fumes emanating from our cars and factories. We propose Utopian communities in gigantic space colonies, but cannot manage our cities. The business world makes us believe that huge industries producing pet foods and cosmetics are a sign of our high standards of living, while economists try to tell us we cannot 'afford' adequate health care, education, or public transport. Medical science and pharmacology are endangering our health, and the Defense Department has become the greatest threat to our national security. Those are the results of overemphasizing our yang, or masculine side—rational knowledge, analysis, expansion—and neglecting our yin, or feminine side—intuitive wisdom, synthesis, and ecological awareness. /26

In a healthy system—an individual, a society, or an ecosystem—there is a balance between integration and self-assertion. This balance is not static but consists of a dynamic interplay between the two complementary tendencies, which makes the whole system flexible and open to change. /27

The overall tone of the book is rather critical, not comforting, and sometimes accusatory. The author makes his point thoroughly, and his lecture is convincing most of the time. What makes the difference to large passages from

The Tao of Physics, in this present book, Capra compares the reality of modern physics, as a largely systemic science, and one that showed there is actually nothing static and 'material' in the configuration of the universe, with our institutional realities, across professional disciplines, to shows us how fragmented and outmoded our policies are, and how we need to reform them along the lines of this new worldview.

Quantum physics clearly shows that the observer is entangled with the object of observation, but we continue to observe natural processes or societal developments in the stern belief that we were 'objective' in any way when we are 'rational minded' and that we could make valid assessments as to the effectiveness of our social and legal policies. In reality, in most areas and professional disciplines, the assessments our leaders are making are incorrect if not mistaken as they are projecting the observer's belief system into the very object of observation. As this is just one of several main characteristics of quantum physics it becomes obvious that the social critique the author offers in this book is multi-faceted, and well-founded. Let us take the example of genetics and see what Fritjof Capra writes:

> Another fallacy of the reductionist approach in genetics is the belief that the character traits of an organism are uniquely determined by its genetic makeup. This 'genetic determinism' is a direct consequence of regarding living organisms as *machines controlled by linear chains of cause and effect*. It ignores the fact that the organisms are multileveled

systems, the genes being imbedded in the chromosomes, the chromosomes functioning within the nuclei of their cells, the cells incorporated in the tissues, and so on. All these levels are involved in mutual interactions that influence the organism's development and result in wide variations of the 'genetic blueprint.' /108 (Italics mine)

Genetics is notoriously one of the most controversial branches of modern science. It is founded on Darwinistic ideas, many of which have been silently disproven in the meantime; in addition a good lot of deterministic thinking is part of this paradigm that believes in the myth of a fate pre-ordained by genes. This has huge and partly dangerous consequences as it is by no means a systemic view but purely reductionist at bottom. The author writes:

> More recently the fallacy of genetic determinism has given rise to a widely discussed theory known as sociobiology, in which all social behavior is seen as predetermined by genetic structure.
>
> Numerous critics have pointed out that this view is not only scientifically unsound but also quite dangerous. It encourages pseudoscientific justifications for *racism and sexism* by interpreting differences in human behavior as genetically preprogrammed and unchangeable. /109 (Italics mine)

The parallels to fascist thinking and ideology are obvious in the whole of the genetics debate, and those who are at the frontline of it, and the myth of the *superman* is lurking for imposing political and social rule over the 'weaker' part of the species. It is certainly not a model that fosters any form of peace and collaboration!

Now let's put a few quotes from this book about modern medicine, our biomedical model, and how it needs to be changed from a mechanistic and fragmented concept to a holistic approach to healing:

> The human body is regarded as a machine that can be analyzed in terms of its parts; disease is seen as the malfunctioning of biological mechanisms which are studied from the point of view of cellular and molecular biology; the doctor's role is to intervene, either physically or chemically, to correct the malfunctioning of a specific mechanism. /118

> The reason for the exclusion of the phenomenon of healing from biomedical science is evident. It is a phenomenon that cannot be understood in reductionist terms. This applies to the healing of wounds, and even more to the healing of illnesses, which generally involve a complex interplay among the physical, psychological, social, and environmental aspects of the human condition. To reincorporate the notion of healing into the theory and practice of medicine, medical science will have to transcend its narrow view of health and illness. This does not mean that it will have to be less scientific. On the contrary, by broadening its conceptual basis it will become more consistent with recent developments in modern science. /119

> The practice of popular medicine has traditionally been the prerogative of women, since the art of healing in the family is usually associated with the tasks and the spirit of motherhood. Folk healers, typically, are both female and male, with proportions varying from culture to culture. They

do not practice within an organized profession but derive their authority from their healing powers—often interpreted as their access to the spirit world—rather than from professional licensing. With the appearance of organized, high-tradition medicine, however, *patriarchal patterns assert themselves and medicine becomes male-dominated.* This is as true for classical Chinese or Greek medicine as for medieval European medicine, or modern cosmopolitan medicine. /121 (Italics mine)

The mechanistic view of the human organism and the resulting engineering approach to health has led to an excessive emphasis on medical technology, which is perceived as the only way to improve health. /147

Hospitals have grown into large professional institutions, emphasizing technology and scientific competence rather than contact with the patient. In these modern medical centers, which look more like airports than therapeutic environments, patients tend to feel helpless and frightened, which often keeps them from getting well. /148

The excessive use of high technology in medical care is not only uneconomic but also causes an unnecessary amount of pain and suffering. Accidents in hospitals now occur more frequently than in any other industries except mining and high-rise construction. /149

Fritjof Capra succeeds to show that the systemic problems we encounter in virtually every branch of the tree of knowledge in our culture are all hanging together and result from a few basic misconceptions that are the result of a myopic mechanistic view, lacking information and training about systemliteracy, the over-evaluation of male values and the general focus upon control, instead of teaching the understanding of collaboration. While nature when health prevails is always harmonious, our society fosters disharmony and extremist thinking in virtually every respect. In our bullet medicine we see doctor-marshals leading war against cancer microbes and using drugs and chemotherapy for fighting a disease that is actually a *shrinking process* of the entire organism, and that has an emotional etiology.

While the truth about disease is that we bring it about ourselves by our unclean ways of thinking and our lacking self-love. Fritjof Capra writes:

> The development of illness involves the continual interplay between physical and mental processes that reinforce one another through a complex network of feedback loops. Disease patterns at any stage appear as manifestations of underlying psychosomatic processes that should be dealt with in the course of therapy. This dynamic view of illness specifically acknowledges the organism's innate tendency to heal itself—to reestablish itself in a balanced state—which may include stages of crises and major life transitions. /364

The author then discusses various alternative healing approaches such as homeopathy or Wilhelm Reich's orgone therapy. As to the discussion of modern alternative cancer cure by the Simontons and others, I have already pointed that out in Chapter Two above. Now to close this extensive review, let me mention a very important detail; it is the more and more prevailing view in biology and psychology that all in fact is energy and that we cannot understand natural processes without considering that all life is after all energy. The author writes:

> As in the new systems biology, the focus of psychology is now shifting from psychological structures to the underlying processes. The human psyche is seen as a dynamic system involving a variety of functions that systems theorists associate with the phenomenon of self-organization. Following Jung and Reich, many psychologists and psychotherapists have come to think of mental dynamics in terms of a flow of energy, and they also believe that these dynamics reflect an intrinsic intelligence—the equivalent of the systems concept of mentation—that enables the psyche not only to create mental illness but also to heal itself. Moreover, inner growth and self-actualization are seen as essential to the dynamics of the human psyche, in full agreement with the emphasis on self-transcendence in the systems view of life. /407

In fact, I have myself undertaken research on the perennial knowledge about the human energy field, and how it integrates today as the 'quantum field,' and over the last years I have witnessed that more and more scientists are working on the integration of this ancient idea into our modern science. It happens both in biology and in psychology. Carl Jung called it the 'psychoenergy' and William A. Tiller coined his own research on the field in the terms of 'psychoenergetic science,' title of one of his books.

Quotes

- These problems (…) are systemic problems, which means that they are closely interconnected and interdependent. They cannot be understood within the fragmented methodology characteristic to our academic disciplines and government agencies. Such an approach will never solve any of our difficulties but will merely shift them around in the complex web of social and ecological relations. A resolution can be found only if the structure of the web is changed, and this will involve profound transformations of our social institutions, values, and ideas. /6

- Studies of periods of cultural transformation in various societies have shown that these transformations are typically preceded by a variety of social indicators, many of them identical to the symptoms of our current crisis. They include a sense of alienation and an increase in mental illness, violent crime, and social disruption, as well as an increased interest in religious cultism—all of which have been observed in our society during the past decade. In times of historic cultural change these indicators have tended to appear one to three decades before the central transformation, rising in frequency and intensity as the transformation is approaching, and falling again after it has occurred. /7

- Ancient Chinese philosophers believed that all manifestations of reality are generated by the dynamic interplay between two polar forces which they called the yin and the yang. Heraclitus, in ancient Greece, compared the world order to an ever living fire, 'kindling in measures and going out in measures.' Empedocles attributed the changes in the universe to the ebb and flow of two complimentary forces, which he called 'love' and 'hate.' (…) After civilizations have reached a peak of vitality, they tend to lose their cultural steam and decline. An essential element in this cultural breakdown, according to Toynbee, is a loss of flexibility. When social structures and behavior patterns have become so rigid that the society can no longer adapt to changing situations, it will be unable to carry on the creative process of cultural evolution. /9

- From the earliest times of Chinese culture, yin was associated with the feminine and yang with the masculine.

This ancient association is extremely difficult to assess today because of its reinterpretation and distortion in subsequent patriarchal eras. In human biology masculine and feminine characteristics are not neatly separated but occur, in varying proportions, in both sexes. Similarly, the Chinese ancients believed that all people, whether men or women, go through yin and yang phases. The personality of each man and each woman is not a static entity but a dynamic phenomenon resulting from the interplay between feminine and masculine elements. This view of human nature is in sharp contrast to that of our patriarchal culture, which has established a rigid order in which all men are supposed to be masculine and all women feminine, and has distorted the meaning of those terms by giving men the leading roles and most of society's privileges. /19

- In view of the original imagery associated with the two archetypal poles, it would seem that yin can be interpreted as corresponding to responsive, consolidating, cooperative activity; yang as referring to aggressive, expanding, competitive activity. Yin action is conscious of the environment, yang action is conscious of the self. In modern terminology one could call the former 'eco-action' and the latter 'ego-action.' /20

- From this it is apparent that rational knowledge is likely to generate self-centered, or yang, activity, where intuitive wisdom is the basis of ecological, or yin, activity. /21

- (…)[I]t is easy to see that our society has consistently favored the yang over the yin—rational knowledge over intuitive wisdom, science over religion, competition over cooperation, exploitation of natural resources over conservation, and so on. (…) According to Chinese wisdom, none of the values pursued by our culture is intrinsically bad, but by isolating them from their polar opposites, by focusing on the yang and investing it with moral virtue and political power, we have brought about the current sad state of affairs. /22

- The view of man as dominating nature and woman, and the belief in the superior role of the rational mind, have been supported and encouraged by the Judeo-Christian tradition, which adheres to the image of a male god, personification of supreme reason and source of ultimate power, who rules the world from above by imposing his divine law on it. The laws of nature searched for by the scientists were seen as reflections of this divine law, originating in the mind of God. /24

- In truth, the understanding of ecosystems is hindered by the very nature of the rational mind. Rational thinking is linear, whereas ecological awareness arises from an intuition of nonlinear systems. One of the most difficult things for people in our culture to understand is the fact that if you do something that is good, then more of the same will not necessarily be better. This, to me, is the essence of ecological thinking. (...) Ecological awareness, then, will arise only when we combine our rational knowledge with an intuition for the nonlinear nature of our environment. Such intuitive wisdom is characteristic of traditional, nonliterate cultures, especially of American Indian cultures, in which life was organized around a highly refined awareness of the environment. (...) Biological evolution of the human species stopped some fifty thousand years ago. From then on, evolution proceeded no longer genetically but socially and culturally, while the human body and brain remained essentially the same in structure and size. /25

- Our progress, then, has been largely a rational and intellectual affair, and this one-sided evolution has now reached a highly alarming stage, a situation so paradoxical that it borders insanity. We can control the soft landings of space craft on distant planets, but we are unable to control the polluting fumes emanating from our cars and factories. We propose Utopian communities in gigantic space colonies, but cannot manage our cities. The business world makes us believe that huge industries producing pet foods and cosmetics are a sign of our high standards of living, while economists try to tell us we cannot 'afford' adequate health care, education, or public transport. Medical science and pharmacology are endangering our health, and the Defense Department has become the greatest threat to our national security. Those are the results of overemphasizing our yang, or masculine side—rational knowledge, analysis, expansion—and neglecting our yin, or feminine side— intuitive wisdom, synthesis, and ecological awareness. /26

- In a healthy system—an individual, a society, or an ecosystem—there is a balance between integration and self-assertion. This balance is not static but consists of a dynamic interplay between the two complementary

- tendencies, which makes the whole system flexible and open to change. /27

- While electromagnetism dethroned Newtonian mechanics as the ultimate theory of natural phenomena, a new trend of thinking arose that went beyond the image of the Newtonian world-machine and was to dominate not only the nineteenth century but all future scientific thinking. /57-58

- In contrast to the mechanistic Cartesian view of the world, the world view emerging from modern physics can be characterized by words like organic, holistic, and ecological. It might also be called a systems view, in the sense of general systems theory. The universe is no longer seen as a machine, made up of a multitude of objects, but has to be pictured as one indivisible dynamic whole whose parts are essentially interrelated and can be understood only as patterns of a cosmic process. /66

- The universe ... is a unified whole that can to some extent be divided into separate parts, into objects made of molecules and atoms, themselves made of particles. But here, at the level of particles, the notion of separate parts breaks down. The subatomic particles—and therefore, ultimately, all parts of the universe, cannot be understood as isolated entities but must be defined through their interrelations. /70

- The crucial feature of quantum theory is that the observer is not only necessary to observe the properties of an atomic phenomenon, but it is necessary even to bring about those properties. My conscious decision about how to observe, say, an electron will determine the electron's properties to some extent. If I ask it a particle question, it will give me a particle answer; if I ask it a wave question, it will give me a wave answer. The electron does not have objective properties independent of my mind. In atomic physics the sharp Cartesian division between mind and matter, between the observer and the observed, can no longer be maintained. We can never speak about nature without, at the same time, speaking about ourselves. /77

- Modern physics thus pictures matter not at all as passive and inert but as being in a continuous dancing and vibrating motion whose rhythmic patterns are determined by the molecular, atomic, and nuclear configurations. We have

THE TURNING POINT

come to realize that there are no static structures in nature. There is stability, but this stability is one of dynamic balance, and the further we penetrate into matter the more we need to understand its dynamic nature to understand its patterns. /79

- All events are interconnected, but the connections are not causal in the classical sense. /80

- The most important consequence of the new relativistic framework has been the realization that mass is nothing but a form of energy. Even an object at rest has energy stored in its mass, and the relation between the two is given by Einstein's famous equation $E = mc^2$, c being the speed of the light. /81

- Subatomic particles must be conceived as four-dimensional entities in space-time. Their forms have to be understood dynamically, as forms in space and time. Particles are dynamic patterns, patterns of activity which have a space aspect and a time aspect. Their space aspect makes them appear as objects with a certain mass, their time aspect as processes involving the equivalent energy. Thus the being of matter and its activity cannot be separated; they are but different aspects of the same space-time reality. (...) In a relativistic description of particle interactions, the forces between the particles—their mutual attraction or repulsion— are pictured as the exchange of other particles. This concept is very difficult to visualize, but it is needed for an understanding of subatomic phenomena. It links the forces between constituents of matter to the properties of other constituents of matter, and thus unifies the two concepts, force and matter, which had seemed to be fundamentally different in Newtonian physics. (...) These energy patterns of the subatomic world form the stable nuclear, atomic, and molecular structures which build up matter and give it its macroscopic solid aspect, thus making us believe that it is made of some material substance. At the macroscopic level this notion of substance is a useful approximation, but at the atomic level it no longer makes sense. Atoms consist of particles, and these particles are not made of any material stuff. When we observe them we never see any substance; what we observe are dynamic patterns continually changing into one another—the continuous dance of energy. /82

- The fact that all the properties of particles are determined by principles closely related to the methods of observation would mean that the basic structures of the material world are determined, ultimately, by the way we look at this world; that the observed patterns of matter are reflections of patterns of the mind. / 85

- From his careful experiments with garden peas, Mendel deduced that there were 'units of heredity'—later to be called genes—that did not blend in the process of reproduction and thus become diluted, but were transmitted from generation to generation without changing their identity. / 107

- Another fallacy of the reductionist approach in genetics is the belief that the character traits of an organism are uniquely determined by its genetic makeup. This 'genetic determinism' is a direct consequence of regarding living organisms as machines controlled by linear chains of cause and effect. It ignores the fact that the organisms are multileveled systems, the genes being imbedded in the chromosomes, the chromosomes functioning within the nuclei of their cells, the cells incorporated in the tissues, and so on. All these levels are involved in mutual interactions that influence the organism's development and result in wide variations of the 'genetic blueprint.' / 108

- More recently the fallacy of genetic determinism has given rise to a widely discussed theory known as sociobiology, in which all social behavior is seen as predetermined by genetic structure. Numerous critics have pointed out that this view is not only scientifically unsound but also quite dangerous. It encourages pseudoscientific justifications for racism and sexism by interpreting differences in human behavior as genetically preprogrammed and unchangeable. / 109

- [The Biomedical Model] The human body is regarded as a machine that can be analyzed in terms of its parts; disease is seen as the malfunctioning of biological mechanisms which are studied from the point of view of cellular and molecular biology; the doctor's role is to intervene, either physically or chemically, to correct the malfunctioning of a specific mechanism. / 118

- The reason for the exclusion of the phenomenon of healing from biomedical science is evident. It is a phenomenon that cannot be understood in reductionist terms. This applies to the healing of wounds, and even more to the healing of illnesses, which generally involve a complex interplay among the physical, psychological, social, and environmental aspects of the human condition. To reincorporate the notion of healing into the theory and practice of medicine, medical science will have to transcend its narrow view of health and illness. This does not mean that it will have to be less scientific. On the contrary, by broadening its conceptual basis it will become more consistent with recent developments in modern science. / 119

- The practice of popular medicine has traditionally been the prerogative of women, since the art of healing in the family is usually associated with the tasks and the spirit of motherhood. Folk healers, typically, are both female and male, with proportions varying from culture to culture. They do not practice within an organized profession but derive their authority from their healing powers—often interpreted as their access to the spirit world—rather than from professional licensing. With the appearance of organized, high-tradition medicine, however, patriarchal patterns assert themselves and medicine becomes male-dominated. This is as true for classical Chinese or Greek medicine as for medieval European medicine, or modern cosmopolitan medicine. / 121

- Rather than trying to understand the psychological dimensions of mental illness, psychiatrists concentrated their efforts on finding organic causes—infections, nutritional deficiencies, brain damage—for all mental disturbances. This 'organic orientation' in psychiatry was furthered by the fact that in several instances researchers could indeed identify organic origins of mental disorders and were able to develop successful methods of treatment. Although these successes were partial and isolated, they established psychiatry firmly as a branch of medicine, committed to the biomedical model. This turned out to be rather a problematic development in the twentieth century. Indeed, even in the nineteenth century the limited success of the biomedical approach to mental illness inspired an alternative movement—the psychological approach—which led to the founding of the dynamic psychiatry and psychotherapy of Sigmund Freud and brought psychiatry much closer to the social sciences and to philosophy. /126-127

- The public image of the human organism—enforced by the content of television programs, and especially by advertising—is that of a machine which is prone to constant failure unless supervised by doctors and treated with medication. The notion of the organism's inherent healing power and tendency to stay healthy is not communicated, and trust in one's own organism is not promoted. (...) It is intriguing and quite ironic that physicians themselves are the ones who suffer most from the mechanistic view of health by disregarding stressful circumstances in their lives. Whereas traditional healers were expected to be healthy people, keeping their body and soul in harmony and in tune with their environment, the typical attitudes and habits of doctors today are quite unhealthy and produce considerable illness. Physicians' life expectancy today is ten to fifteen years less than that of the average population, and they have not only high rates of physical illness but also high rates of alcoholism, drug abuse, suicide, and other social pathologies. /146

- The mechanistic view of the human organism and the resulting engineering approach to health has led to an excessive emphasis on medical technology, which is perceived as the only way to improve health. /147

- Hospitals have grown into large professional institutions, emphasizing technology and scientific competence rather than contact with the patient. In these modern medical centers, which look more like airports than therapeutic environments, patients tend to feel helpless and frightened, which often keeps them from getting well. /148

- The excessive use of high technology in medical care is not only uneconomic but also causes an unnecessary amount of pain and suffering. Accidents in hospitals now occur more frequently than in any other industries except mining and high-rise construction. /149

- The theory of specific disease causation has been successful in a few special cases, such as acute infectious processes and nutritional deficiencies, but the overwhelming majority of illnesses cannot be understood in terms of the reductionist concepts of well-defined disease entities and single causes. The main error of the biomedical approach is the confusion between disease processes and disease origins. Instead of asking why an illness occurs, and trying to remove the conditions that lead to it, medical researchers try to understand the biological mechanisms through which the disease operates, so that they can then interfere with them. /150

- The origins of disease will generally be found in several causative factors that must concur to result in ill health. Moreover, their effects will differ profoundly from person to person, since they depend on the individual's emotional reactions to stressful situations and on the social environment in which these situations occur. /151

- Whereas illness is a condition of the total human being, disease is a condition of a particular part of the body, and rather than treating patients who are ill, doctors have concentrated on treating their diseases. /152

- Perhaps the most striking example of the emphasis on symptoms rather than underlying causes is the drug approach of contemporary medicine. It has its roots in the erroneous view that bacteria are the primary causes of disease, rather than symptomatic manifestations of underlying physiological disorder. /153

- There seem to be very few infectious diseases in which the bacteria cause actual direct damage to the cells or tissues of the host organism. There are some, but in most cases the damage is caused by an overreaction of the organism, a kind of panic in which a number of powerful, unrelated defense mechanisms are all turned on at once. Infectious diseases, then, arise most of the time from a lack of coordination within the organism, rather than from injury caused by invading bacteria. /155

- An important aspect of the mechanistic view of living organisms and the resulting engineering approach to health is the belief that the cure of illness requires some outside intervention by the physician, which can be either physical, through surgery or radiation, or chemical, through drugs. /157

- The systems view looks at the world in terms of relationships and integration. Systems are integrated wholes whose properties cannot be reduced to those of smaller units. Instead of concentrating on basic building blocks or basic substances, the systems approach emphasizes basic principles of organization. /286

- Systems thinking is process thinking; form becomes associated with process, interrelation with interaction, and opposites are unified through oscillation. /288

- Machines function according to linear chains of cause and effect, and when they break down a single cause for the breakdown can usually be identified. In contrast, the functioning of organisms is guided by cyclical patterns of information flow known as feedback loops. /289

- This nonlinear interconnectedness of living organisms indicates that the conventional attempts of biomedical science to associate diseases with single causes are highly problematic. Moreover, it shows the fallacy of 'genetic determination,' the belief that various physical or mental features of an individual organism are 'controlled' or 'dictated' by its genetic makeup. The systems view makes it clear that genes do not uniquely determine the functioning of an organism as cogs and wheels determine the working of a clock. Rather, genes are integral parts of an ordered whole and thus conform to its systemic organization. /289-290

- Living organisms function quite differently. They are open systems, which means that they have to maintain a continuous exchange of energy and matter with their environment to stay alive. This exchange involves taking in ordered structures, such as food, breaking them down and using some of their components to maintain or even increase the order of the organism. This process is known as metabolism. It allows the system to remain in a state of nonequilibrium, in which it is always 'at work.' A high degree of nonequilibrium is absolutely necessary for self-organization; living organisms are open systems that continually operate far from equilibrium. /291

- The stability of self-organizing systems is utterly dynamic and must not be confused with equilibrium. It consists in maintaining the same overall structure in spite of ongoing changes and replacements of its components. (…) We replace all our cells, except for those in the brain, within a few years, yet we have no trouble recognizing our friends even after long periods of separation. Such is the dynamic stability of self-organizing systems. /292

- Self-renewal is an essential aspect of self-organizing systems. Whereas a machine is constructed to produce a specific product or to carry out a specific task intended by its designer, an organism is primarily engaged in renewing itself; cells are breaking down and building up structures, tissues and organs are replacing their cells in continual cycles. Thus the pancreas replaces most of its cells every twenty-four hours, the stomach lining every three days; our white blood cells are renewed in ten days and 98 percent of the protein in the brain is turned over in less than one month. All these processes are regulated in such a way that the overall pattern of the organism is preserved, and this remarkable ability of self-maintenance persists under a variety of circumstances, including changing environmental conditions and many kinds of interference. /293

- Such as state is known as homeostasis. It is a state of dynamic, transactional balance in which there is great flexibility; in other words, the system has a large number of options for interacting with its environment. When there is some disturbance, the organism tends to return to its original state, and it does so by adapting in various ways to environmental changes. Feedback mechanisms come into play and tend to reduce any deviation from the balanced state. /294

- The more one studies the living world the more one comes to realize that the tendency to associate, establish links, live inside one another and cooperate is an essential characteristic of living organisms. /301

- Excessive aggression, competition, and destructive behavior are predominant only in the human species and have to be dealt with in terms of cultural values rather than being 'explained' pseudoscientifically as inherently natural phenomena. /302

- The tendency of living systems to form multileveled structures whose levels differ in their complexity is all-pervasive throughout nature and has to be seen as a basic principle of self-organization. At each level of complexity we encounter systems that are integrated, self-organizing wholes consisting of smaller parts and, at the same time, acting as parts of larger wholes. For example, the human organism contains organ systems composed of several organs, each organ being made up of tissues and each tissue made up of cells. The relations between these systems levels can be represented by a 'systems tree.' (…) Thus, the pervasiveness of order in the universe takes on a new meaning: order at one systems level is the consequence of self-organization at a larger level. / 303

- This creative reaching out into novelty, which in time leads to an ordered unfolding of complexity, seems to be a fundamental property of life, a basic characteristic of the universe which is not—at least for the time being—amenable to further explanation. We can, however, explore the dynamics and mechanisms of self-transcendence in the evolution of individuals, species, ecosystems, societies, and cultures. / 309

- Evolution is an ongoing and open adventure that continually creates its own purpose in a process whose detailed outcome is inherently unpredictable. Nevertheless, the general pattern of evolution can be recognized and is quite comprehensible. Its characteristics include the progressive increase of complexity, coordination, and interdependence; the integration of individuals into multileveled systems; and the continual refinement of certain functions and patterns of behavior. / 313

- There are larger manifestations of mind of which our individual minds are only subsystems. This recognition has very radical implications for our interactions with the natural environment. If we separate mental phenomena from the larger systems in which they are immanent and confine them to human individuals, we will see the environment as mindless and will tend to exploit it. Our attitudes will be very different when we realize that the environment is not only alive but also mindful, like ourselves. / 316

- Our responses to the environment, then, are determined not so much by the direct effect of external stimuli on our biological system but rather by our past experience, our expectations, our purposes, and the individual symbolic interpretation of our perceptual experience. / 321

- In the future elaboration of the new holistic world view, the notion of rhythm is likely to play a very fundamental role. The systems approach has shown that living organisms are intrinsically dynamic, their visible forms being stable manifestations of underlying processes. Process and stability, however, are compatible only if the processes form rhythmic patterns—fluctuations, oscillations, vibrations, waves. The new systems biology shows that fluctuations are crucial in the dynamics of self-organization. They are the basis of order in the living world: ordered structures arise from rhythmic patterns. / 326-327

- The conceptual shift from structure to rhythm may be extremely useful in our attempts to find a unifying description of nature. / 327

- The notion of illness as originating in a lack of integration seems to be especially relevant to approaches that try to understand living organisms in terms of rhythmic patterns. From this perspective synchrony becomes an important measure of health. Individual organisms interact and communicate with one another by synchronizing their rhythms and thus integrating themselves into the larger rhythms of their environment. / 355

- When the systems view of mind is adopted, it becomes obvious that any illness has mental aspects. Getting sick and healing are both integral parts of an organism's self-organization, and since mind represents the dynamics of this self-organization, the processes of getting sick and of healing are essential mental phenomena. Because mentation is a multileveled pattern of processes, most of them are taking place in the unconscious realm, we are not always fully aware of how we move in and out of illness, but this does not alter the fact that illness is a mental phenomenon in its very essence. / 359

- The precise ways in which physical and psychological patterns interlink are still little understood, and thus most physicians tend to restrict themselves to the biomedical model and neglect the psychological aspects of illness. However, there have been significant attempts to develop a unified approach to the mind/body system throughout the history of Western medicine. Several decades ago these attempts culminated in the foundation of psychosomatic medicine as a scientific discipline, concerned specifically with the study of the relationships between the biological and psychological aspects of health.* This new branch of medicine is now rapidly gaining acceptance, especially with the growing awareness of the relevance of stress, and it is likely to play an important role in a future holistic system of health care. /359 (*Referencing Lipowski, Z.J. 1977 'Psychosomatic Medicine in the Seventies: An Overview' The American Journal of Psychiatry. March).

- The first step in this kind of self-healing will be the patients' recognition that they have participated consciously or unconsciously in the origin and development of their illness, and hence will also be able to participate in the healing process. In practice, this notion of patient participation, which implies the idea of patient responsibility, is extremely problematic and is vigorously denied by most patients. Conditioned as they are by the Cartesian framework, they refuse to consider the possibility that they may have participated in their illness, associating the idea with blame and moral judgment. It will be important to clarify exactly what is meant by patient participation and responsibility. /361

- Mental attitudes and psychological techniques are important means for both the prevention and healing of illness. A positive attitude combined with specific stress-reduction techniques will have a strong positive impact on the mind/body system and will often be able to reverse the disease process, even to heal severe biological disorders. The same techniques can be used to prevent illness by applying them to cope with excessive stress before any serious damage occurs. An impressive proof of the healing power of positive expectations alone is provided by the placebo effect. A placebo is an imitation medicine, dressed up like an authentic pill and given to patients who think they are receiving the real thing. Studies have shown that 35 percent of patients consistently experience 'satisfactory relief' when placebos are used instead of regular medication for a wide range of medical problems. Placebos have been strikingly successful in reducing or eliminating physical symptoms, and have produced dramatic recoveries from illnesses for which there are no known medical cures. The only active ingredient in these treatments appears to be the power of the patient's positive expectations, supported by interaction with the therapist. /362 (Referencing Cousins, Norman. 1977. 'The Mysterious Placebo.' Saturday Review, October 1).

- In the past, psychosomatic self-healing has always been associated with faith in some treatment—a drug, the power of a healer, perhaps a miracle. In a future approach to health and healing, based on the new holistic paradigm, it should be possible to acknowledge the individual's potential for self-healing directly, with no need for any conceptual crutches, and to develop psychological techniques that will facilitate the healing process. /363

- The development of illness involves the continual interplay between physical and mental processes that reinforce one another through a complex network of feedback loops. Disease patterns at any stage appear as manifestations of underlying psychosomatic processes that should be dealt with in the course of therapy. This dynamic view of illness specifically acknowledges the organism's innate tendency to heal itself—to reestablish itself in a balanced state—which may include stages of crises and major life transitions. /364

- The reorganization of health care will also mean discouraging the construction and use of facilities that are inefficient and incompatible with the new view of health. To change the present technology-intensive, hospital-based system, a first useful step may be, as Victor Fuchs has suggested, to impose a moratorium on all hospital construction and expansion to bring our escalating hospital costs under control. At/ the same time hospitals will gradually be transformed into more efficient and more humane institutions, comfortable and therapeutic environments modeled after hotels rather than factories or machine shops, with good and nourishing food, family members included in patient care, and other such sensible improvements. /370-371 (Referencing White, Kerr L. 1978. 'Ill Health and Its Amelioration: Individual and Collective Choices.' In Carlson, Rick, J. ed. Future Directions in Health Care: A New Public Policy, Cambridge, Mass.: Ballinger, as well as Fuchs, Victor R. 1974. Who Shall Live? New York: Basic Books).

- A number of therapeutic models and techniques are already being developed that go beyond the biomedical framework and are consistent with the systems view of health. Some of them are based on well-established Western healing traditions, others are of more recent origin, and most of them are not taken very seriously by the medical establishment because they are difficult to understand in terms of classical scientific concepts.

To begin with, numerous unorthodox approaches to health share a belief in the existence of patterns of 'subtle energies,' or 'life philosophies,' and see illness as resulting from changes in these patterns. Although the therapies practiced in these traditions, which are sometimes referred to as 'energy medicine,' involve a variety of techniques, all of them are believed to influence the organism at a more fundamental level than the physical or psychological symptoms of illness. This view is quite similar to that of the Chinese medical tradition, and so are many of the concepts used in the various healing traditions. For example, when homeopaths speak about the 'vital force,' or Reichian therapists about 'bioenergy,' they use these terms in a sense that comes very close to the Chinese concept of ch'i. /373

- Homeopathic therapy consists of matching the pattern of symptoms that is characteristic of the patient with a similar pattern characteristic of the remedy. Vithoulkas believes that each remedy is associated with a certain vibrational pattern that constitutes its very essence. When the remedy is taken its energy pattern resonates with the energy pattern of the patient and thereby induces the healing process. The resonance phenomenon seems to be central to homeopathic therapy, / but what exactly resonates and how this resonance is brought about is not well understood. /375-376

- Homeopathic remedies are substances derived from animals, plants and minerals, and are taken in highly diluted form. The selection of the correct remedy is based on Hahnemann's Law of Similars—'Like Cures Like'—which gave homeopathy its name. According to Hahnemann, any substance that can produce a total pattern of symptoms in a healthy human being can cure those same symptoms in a sick person. Homeopaths claim that literally any substance can produce, and cure, a wide spectrum of highly individualized symptoms known as the 'personality' of the remedy. /376

- From the very beginning of his medical research, Reich was keenly interested in the role of energy in the functioning of living organisms, and one of the main goals of his psychoanalytic work was to associate the sexual drive, or libido, which Freud saw as an abstract psychological force, with concrete energy flowing through the physical organism. This approach led Reich to the concept of bioenergy, a fundamental form of energy that permeates and governs the entire organism and manifests itself in the emotions as well as in the flow of bodily fluids and other biophysical movements. Bioenergy, according to Reich, flows in wave movements and its basic dynamic characteristic is pulsation. Every mobilization of flow processes / and emotions in the organism is based on a mobilization of bioenergy. /377-378

- One of Reich's key discoveries was that attitudes and emotional experiences can give rise to certain muscular patterns which block the free flow of energy. These muscular blocks, which Reich called 'character armor,' are developed in nearly every adult individual. They reflect our personality and enclose key elements of our emotional history, locked up in the structure and tissue of our muscles. The central task of Reichian therapy is to destroy the muscular armor in order to reestablish the organism's full capacity for the pulsation of bioenergy. /378

- It is evident that Reich's concept of bioenergy comes very close to the Chinese concept of ch'i. Like the Chinese, Reich emphasized the cyclical nature of the organism's flow processes and, like the Chinese, he also saw the energy flow in the body as the reflection of a process that goes on in the universe at large. To him bioenergy was a special manifestation of a form of cosmic energy that he called 'orgone energy.' Reich saw this orgone research as some kind of primordial substance, present everywhere in the atmosphere and extending through all space, like the ether of the nineteenth-century physics. Inanimate as well as living matter, according to Reich, derives from orgone energy through a complicated process of differentiation. /378

- Influenced by the pioneering ideas of Wilhelm Reich, by Eastern concepts, and by the modern dance movement, a number of therapists have combined various elements from these traditions to develop bodywork techniques that have recently become very popular. The principal founders of these new approaches are Alexander Lowen ('bioenergetics'), Frederick Alexander ('Alexander technique'), Moshe Feldenkrais ('functional integration'), Ida Rolf ('structuring integration'), and Judith Aston ('structural patterning'). In addition, various massage therapies have been developed, many of them inspired by Eastern techniques like shiatsu and acupressure. All these approaches are based on the Reichian notion that emotional stress manifests itself in the form of blocks in the muscle structure and tissue, but they differ in the methods employed for releasing these psychosomatic blocks. (Referencing Popenoe, Cris. 1977. Wellness. Washington, D.C.).

- The imbalance and fragmentation that pervade our culture play an important role in the development of cancer and, at the same time, prevent medical researchers and clinicians from understanding the disease or treating it successfully. The conceptual framework and therapy that Carl Simonton, a radiation oncologist, and Stephanie Matthews-Simonton, a psychotherapist, have developed are fully consistent with the views of health and illness we have been discussing and have far-reaching implications for many areas of health and healing. (Reference: *Oncology*, from the Greek onkos ('mass'), is the study of tumors, and Simonton, O. Carl, Matthews-Simonton, Stephanie, and Creighton, James, 1978. Getting Well Again. Los Angeles: Tarcher).

- The popular image of cancer has been conditioned by the fragmented world view of our culture, the reductionist approach of our science, and technology-oriented practice of medicine. Cancer is seen as a strong and powerful invader / that strikes the body from outside. There seems to be no hope of controlling it, and for most people cancer is synonymous with death. Medical treatment—whether radiation, chemotherapy, surgery, or a combination of these—is drastic, negative, and further injures the body. Physicians are increasingly coming to see cancer as a systemic disorder; a disease that has a localized appearance but has the ability to spread, and that really involves the entire body, the original tumor being merely the tip of the iceberg. / 388-389

- One of the main aims of the Simonton approach is to reverse the popular image of cancer, which does not correspond to the findings of current research. Modern cellular biology has shown that cancer cells are not strong and powerful but, on the contrary, weak and confused. They do not invade, attack, or destroy, but simply overproduce. A cancer begins with a cell that contains incorrect genetic information because it has been damaged by harmful substances or other environmental influences, or simply because the organism will occasionally produce an imperfect cell. The faulty information will prevent the cell from functioning normally, and if this cell reproduces others with the same incorrect genetic makeup, the result will be a tumor composed of a mass of these imperfect cells. Whereas normal cells communicate effectively with their environment to determine their optimal size and rate of reproduction, the communication and self-organization of malignant cells are impaired. As a result they grow larger than healthy cells and reproduce recklessly. Moreover, the normal cohesion between cells may weaken and malignant cells may break loose from the original mass and travel to other parts of the body to form new tumors—which is known as metastasis. In a healthy organism the immune system will recognize abnormal cells and destroy them, or at least wall them off so they cannot spread. But if for some reason the immune system is not strong enough, the mass of faulty cells will continue to grow. Cancer, then, is not an attack from without but a breakdown within. / 389-390

> The Simontons and other researchers have developed a psychosomatic model of cancer that shows how psychological and physical states work together in the onset of the disease. Although many details of this process still need to be clarified, it has become clear that the emotional stress has two principal effects. It suppresses the body's immune system and, at the same time, leads to hormonal imbalances that result in an increased production of abnormal cells. Thus optimal conditions for cancer growth are created. The production of malignant cells is enhanced precisely at a time when the body is least capable of destroying them.
>   As far as the personality configuration is concerned, the individual's emotional states seem to be the crucial element in the development of cancer. The connection between cancer and emotions has been observed for hundreds of years, and today there is substantial evidence for the significance of specific emotional states. These are the result of a particular life history that seems to be characteristic of cancer patients. Psychological profiles of such patients have been established by a number of researchers, some of whom were even able to predict the incidence of cancer with remarkable accuracy on the basis of these profiles. / 391

- As in any holistic therapy, the first step toward initiating the healing cycle consists of making patients aware of the wider context of their illness. Establishing the context of cancer begins by asking patients to identify the major stresses occurring in their lives six to eighteen months prior to their diagnosis. The list of these stresses is then used as a basis for discussing the patients' participation in the onset of their disease. The purpose of the concept of patient participation is not to evoke guilt, but rather to create the basis for reversing the cycle of psychosomatic processes that led to the state of ill health. /392

- While the Simontons are establishing the context of a patient's illness, they are also strengthening his belief in the effectiveness of the treatment and the potency of the body's defenses. The development of such a positive attitude is crucial for the treatment. Studies have shown that the patient's response to treatment depends more on his attitude than on the severity of the disease. Once feelings of hope and anticipation are generated, the organism translates them into biological processes that begin to restore balance and to revitalize the immune system, using the same pathways that were / used in the development of the illness. /392-393

- The Simontons see cancer not merely as a physical problem but as a problem of the whole person. Accordingly, their therapy does not focus on the disease alone but deals with the total human being. It is a multidimensional approach involving various treatment strategies designed to initiate and support the psychosomatic process of healing. At the biological level the aim is twofold; to destroy cancer cells and to revitalize the immune system. In addition, regular physical exercise is used to reduce stress, to alleviate depression, and to help patients get more in touch with their bodies. Experience has shown that cancer patients are capable of far more physical activity than most people would assume. /393

- One of the most vital and enthusiastic movements that arose from dissatisfaction with the mechanistic orientation of psychological thought is the school of humanistic psychology spearheaded by Abraham Maslow. Maslow rejected Freud's view of humanity as being dominated by lower instincts and criticized Freud for deriving his theories of human behavior from the study of neurotic and psychotic individuals. Maslow thought that conclusion based on observing the worst in human beings rather than the best were bound to result in a distorted view of human nature. /402

- The healthy experience of oneself is an experience of the whole organism, body and mind, and mental illnesses often arise from failure to integrate the various components of this organism. From this point of view the Cartesian split between mind and body and the conceptual separation of individuals from their environment appear to be symptoms of a collective mental illness shared by most of Western culture, as they are indeed often perceived by other cultures. /406

- As in the new systems biology, the focus of psychology is now shifting from psychological structures to the underlying processes. The human psyche is seen as a dynamic system involving a variety of functions that systems theorists associate with the phenomenon of self-organization. Following Jung and Reich, many psychologists and psychotherapists have come to think of mental dynamics in terms of a flow of energy, and they also believe that these dynamics reflect an intrinsic intelligence—the equivalent of the systems concept of mentation—that enables the psyche not only to create mental illness but also to heal itself. Moreover, inner growth and self-actualization are seen as essential to the dynamics of the human psyche, in full agreement with the emphasis on self-transcendence in the systems view of life. /407

Chapter Seven

Uncommon Wisdom

Uncommon Wisdom

Conversations with Remarkable People
New York: Bantam Books, 1989

Review

Uncommon Wisdom is not a science book in the strict sense, but it elucidates much about the scientist Fritjof Capra and the method of his special approach to knowledge gathering by exchanging views with others for achieving a multi-vectorial perspective. It is autobiographic also in this respect, of course, and that has value in itself, for until now Fritjof Capra has not agreed to be biographed, as I learnt from an employee at the Berkeley Center of Ecoliteracy just recently.

It is a very readable and from the human point of view highly interesting book, for it shows with many examples that we arrive at a mature judgment of any problem only by exchanging with others, and if the field of study is outside our professional expertise, by consulting with the best experts in the field.

I reviewed *Uncommon Wisdom (1989)* only recently, and after my second lecture of the book. Previously, I had been convinced that the book cannot be reviewed as it is very personal, autobiographic and contains many conversations difficult if not impossible to paraphrase without actually quoting them. To quote them entirely was excluded because of copyright, so I had to mark the main points only.

First of all, I reflected why I should review the book. After my initial hesitation, and reading it once again, I came to realize that it is actually a very important document, because it relates the transition that the author made from *The Tao of Physics (1975)* to *The Turning Point (1987)*, and how Capra was receiving broad feedback and support from other scientists and professionals, psychologists, psychiatrists and medical doctors to discuss his paradigm-changing research, and the project for the upcoming book that was certainly challenging to write. As such, the book is something like a background study for Capra's upcoming bestseller *The Turning Point (1987)* while it was published two years aft er the latter.

The book contains conversations with *Werner Heisenberg, J. Krishnamurti, Geoffrey Chew, Gregory Bateson, Stanislav Grof, R.D. Laing, Carl Simonton, Margaret Lock, E.F. Schumacher, Hazel Henderson,* and *Indira Gandhi.* In addition, the so-called *Big Sur Dialogues*, a conversation about paradigm changes in medicine, at the *Esalen Institute*, which was led by Capra, and to which attended and contributed *Gregory Bateson, Antonio Dimalanta, Stanislav Grof, Hazel Henderson, Margaret Lock, Leonard Shlain* and *Carl Simonton.*

Uncommon Wisdom is certainly a must-read for everyone who wants to be informed how, since more than two decades, our fundamental scientific paradigms are changing toward a holistic and systemic worldview.

All the scholars Fritjof Capra met, and other people he mentioned in this book do not need to be introduced, as they are world-famous.

To begin with, I could not say which part of the book I liked best and which part, as it is often the case, was of lesser interest to me. It was all one fascinating read from the first to the last word. Perhaps, yes, the most captivating accounts for me were Capra's meetings with *Gregory Bateson, Stanislav Grof* and *Ronald David Laing*. This is by the way my experience with all of Capra's books, and I believe this has to do with both his scientific honesty and his clear, and careful writing style that doesn't venture into speculations, but still conveys the *emotional nature* of the author.

Capra is perhaps exceptional among scientists in that respect, and in this book this becomes particularly evident, as it retraces also his hippie years, and his spirit of adventure as a young man, lover, father, artist and scientist.

What emerges from the lecture of this book is a deep insight not only in the scientific subjects discussed in it, but in the way Capra researches.

As he has outlined it in his lecture at Grace Cathedral, San Francisco, in November 2007, his research method is unique in that he doesn't as other researchers base his knowledge-gathering on books, as the primary source of information, but on their authors. Over the many years of his research and publishing, he managed to always get in touch with the authors of the books he found important for his research, and bonds with them, and often actually befriends them. Sometimes, he spontaneously sent a manuscript to some of them, and received valuable feedback. In this way, Fritjof Capra has befriended many great minds over the last thirty years, among them those featured in this fascinating and very personal book.

Let me start this review with how Capra described the hippie movement and his involvement in the counterculture. Many of his observations are highly accurate; they are of course also revealing as to his personality and how he actually revolted against 'the system' at that time, and was living out his rebellious nature:

> The hippies opposed many cultural traits that we, too, found highly unattractive. To distinguish themselves from the crew cuts and polyester suits of the straight business executives they wore long hair, colorful and individualistic clothes, flowers, beads, and other jewelry. They lived naturally without disinfectants or deodorants, many of them vegetarians, many practicing yoga or some other form of meditation. They would often bake their own bread or practice some craft. They were called 'dirty hippies' by the straight society but referred to themselves as 'the beautiful people.' Dissatisfied with a system of education that was designed to prepare young people for a society they rejected, many hippies dropped out of the educational system even though they were often highly talented. This subculture was immediately identifiable and tightly bound together. It had its own rituals, its music, poetry, and literature, a common fascination with spirituality and the occult, and the shared vision of a peaceful and beautiful society. Rock music and psychedelic drugs were powerful bonds that strongly influenced the art and life-style of the hippie culture. /23

Now, how did this counterculture impact upon Fritjof Capra's worldview and his personal development as a scientist who more and more developed a critical view of our science tradition?

> The sixties brought me without doubt the deepest and most radical personal experiences of my life: the rejection of conventional, 'straight' values; the closeness, peacefulness, and trust of the hippie community; the freedom of communal nudity; the expansion of consciousness through psychedelics and meditation; the playfulness and attention to the 'here and now'—all of which resulted in a continual sense of magic, awe, and wonder that, for me, will forever be associated with the sixties. / 24

> The sixties were also the time when my political consciousness was raised. This happened first in Paris, where many graduate students and young research fellows were also active in the student movement that culminated in the memorable revolt that is still known simply as 'May 68.' I remember long discussions at the Science Faculty at Orsay, during which the students not only analyzed the Vietnam War and the Arab-Israeli war of 1967, but also questioned the power structure within the university and discussed alternative, nonhierarchical structures. / Id.

> I remember that my sympathy with the Black Power movement was aroused by a dramatic and unforgettable event shorty after we moved to Santa Cruz. We read in the newspaper that an unarmed black teenager had been brutally shot to death by a white policeman in a small record store in San Francisco. Outraged, my wife and I drove to San Francisco and went to the boy's funeral, expecting to find a large crowd of like-minded white people. Indeed, there was

a large crowd, but to our great shock we found that, together with another two or three, we were the only whites. The congregation hall was lined with fierce-looking Black Panthers clad in black leather, arms crossed. The atmosphere was tense and we felt insecure and frightened. But when I approached one of the guards and asked whether it would be all right for us to attend the funeral, he looked straight into my eyes and said simply, 'You're welcome, brother, you're welcome!' /25

In the following, Capra explains how he felt meeting with great philosophical scholars of his time, such as Alan Watts, Carlos Castaneda or J. Krishnamurti, and what Zen meant for him personally.

After moving to California, I soon found out that Alan Watts was one of the heroes of the counterculture, whose books were on the shelves of most hippie communes, along with those of Carlos Castaneda, J. Krishnamurti, and Hermann Hesse. Although I had read books about Eastern philosophy and religion before reading Watts, it was he who helped me most to understand its essence. His books would take me as far as one could go with books and would stimulate me to go further through direct, nonverbal experience. /26

The impact of Krishnamurti's physical appearance and charisma was enhanced and deepened by what he said. Krishnamurti was a very original thinker who rejected all spiritual authority and traditions. His teachings were quite close to those of Buddhism, but he never used any terms from Buddhism or from any other branch of traditional Eastern thought. Tee task he set himself was extremely difficult—to use language and reasoning in order to lead his audience beyond language and reasoning—and the way in which he went about it was highly impressive. / 28

The following quote reminds us of *The Tao of Physics* and my critique of it. I can really not see how Krishnamurti could have solved a problem for him that was based upon a problematic conceptual approach? Krishnamurti merely told him that he could define himself as a scientist without however needing to limit his conceptual outlook to what was accepted only by the scientific method. Krishnamurti meant that he could see himself as a human being, without the limitative views of a scientist but I doubt if K wanted to justify his approach to grossly compare mysticism with science, thereby defending a quite nonsensical proposition that can be placed neither in the realm of science nor in the realm of philosophy!

> The problem that Krishnamurti had solved for me, Zen-like with one stroke, is the problem most physicists face when confronted with the ideas of mystical traditions—how can one transcend thinking without losing one's commitment to science? It is the reason, I believe, that so many of my colleagues feel threatened by my comparisons between physics and mysticism. / 31

I do not believe that Capra's scientific colleagues really felt 'threatened' by his comparisons between physics and mysticism. I tend to believe they felt that such comparisons are either utterly personal and thus autobiographic without having a meaning for others, or they are nonsensical and thus will estrange a scientist from a conceptual point of view. They may without further have had a good place in the hippie communities and with new agers, but it is no wonder that they were generally unacceptable for most scientists—and for good reason! Zen is of course a mental training that everyone can profit from but why should it have other than random experiential value for a scientist?

> The Zen tradition, in particular, developed a system of nonverbal instruction through seemingly nonsensical riddles, called koans, which cannot be solved by thinking. They are designed precisely to stop the thought process and thus make the student ready for the nonverbal experience of reality. /32

> When I first read about the koan method in Zen training, it had a strangely familiar ring to me. I had spent many years studying another kind of paradox that seemed to play a similar role in the training of physicists. There were differences, of course. My own training as a physicist certainly had not had the intensity of Zen training. But then I thought about Heisenberg's account of the way in which physicists in the 1920s experienced the quantum paradoxes, struggling for understanding in a situation where nature alone was the teacher. The parallel was obvious and fascinating and, later on, when I learned more about Zen Buddhism, I found that it was indeed very significant. As in Zen, the solutions to the physicist's problems were hidden in paradoxes that could not be solved by logical reasoning but had to be understood in terms of a new awareness, the awareness of the atomic reality. Nature was their teacher and, like the Zen masters, she did not provide any statements; she just provided the riddles. /32

An interesting detail about his writing experience of *The Tao of Physics* is reported here:

> Shortly before leaving California I had designed a photo-montage—a dancing Shiva superimposed on tracks of colliding particles in a bubble chamber—to illustrate my experience of the cosmic dance on the beach. One day I sat in my tiny room near Imperial College and looked at this beautiful picture, and suddenly I had a very clear realization. I knew with absolute certainty that the parallels between physics and mysticism, which I had just begun to discover, would someday be common knowledge; I also knew that I was best placed to explore these parallels thoroughly and to write a book about them. I resolved there and then to write that book, but I also decided that I was not yet ready to do so. I would first study my subject further and write a few articles about it before attempting the book. /35

For Fritjof Capra, as for many of us today, including myself, *Taoism* and the teaching of Lao-tzu was an important philosophy that had a strong impact also upon his scientific worldview:

> The Taoist sages concentrated their attention fully on the observation of nature in order to discern the 'characteristics of the Tao.' In doing so they developed an attitude that was essentially scientific; only their deep mistrust of the analytic method of reasoning prevented them from constructing proper scientific theories. Nevertheless, their careful observation of nature, combined with a strong mystical intuition, led them to profound insights which are confirmed by modern scientific theories. The deep ecological wisdom, the empirical approach, and the special flavor of Taoism,

which I can best describe as 'quiet ecstasy,' were enormously attractive to me, and so Taoism quite naturally became the way for me to follow. /36

In the following Dr. Capra describes how the writings of Carlos Castaneda mattered to him, as well as Buddhism:

Castaneda, too, exerted a strong influence on me in those years, and his books showed me yet another approach to the spiritual teachings of the East. I found the teachings of the American Indian traditions, expressed by the legendary Yaqui sage Don Juan, very close to those of the Taoist tradition, transmitted by the legendary sages Lao Tzu and Chuang Tzu. The awareness of being embedded in the natural flow of things and the skill to act accordingly are central to both traditions. As the / Taoist sage flows in the current of the Tao, the Yaqui 'man of knowledge' needs to be light and fluid to 'see' the essential nature of things. /36-37

The strongest influence of Buddhist tradition on my own thinking has been the emphasis on the central role of compassion in the attainment of knowledge. According to the Buddhist view, there can be no wisdom without compassion, which means for me that science is of no value unless it is accompanied by social concern. /37

More autobiographical details about his life as a young scientist and writer are revealed in these quotes:

Although the years 1971 and 1972 were very difficult for me, they also were very exciting. I continued my life as part-time physicist and part-time hippie, doing research in particle physics at Imperial College while also pursuing my larger research in an organized and systematic way. I managed to get several part-time jobs —teaching high-energy physics to a group of engineers, translating technical texts from English into German, teaching mathematics to high school girls—which made enough money for me to survive but did not allow for any material luxury. My life during those two years was very much like that of a pilgrim; its luxuries and joys were not those of the material plane.

What carried me through this period was a strong belief in my vision and a conviction that my persistence would eventually be rewarded. During those two years I always had a quote from the Taoist sage Chuang Tzu pinned to my wall: 'I have sought a ruler who would employ me for a long time. That I have not found one shows the character of the time.' /37

I packed my suit, shirts, leather shoes, and physics papers in a bag, put on my patched jeans, sandals, and flowered shirt, and hit the road. The weather was superb and I greatly enjoyed traveling through Europe the slow way, meeting lots of people and visiting beautiful old towns on the way. My overriding experience on this trip, the first to Europe after two years of California, was the realization that European national borders are rather artificial divisions. I noticed that the language, customs, and physical characteristics of the people did not change abruptly at the borders, but rather gradually, and that the people on either side of the border often had much more in common with each other than, say, with the inhabitants of the capitals of their countries. /38

Here are some difficulties reported about *The Tao* and the publishing process that the reader may never have heard about:

Today *The Tao of Physics* is an international bestseller and is often praised as a classic that has influenced many other writers. But when I planned to write it, it was extremely difficult for me to find a publisher.

Friends in London who were writers suggested that I should first look for a literary agent, and even that took considerable time. When I finally found an agent who agreed to take on this unusual project, he told me that he would need an outline of the book plus three sample chapters to offer to prospective publishers. This put me in a great dilemma. /45

In the meantime, my agent offered the manuscript to the major publishers in London and New York, all of / whom turned it down. After a dozen rejections, a small but enterprising London publishing firm, Wildwood House, accepted the proposal and paid me an advance that gave me sufficient support to write the entire book. /45-46

Interesting developments follow about Jeffrey Chew's bootstrap approach to modern physics, which are very interesting and informative:

> According to the bootstrap hypothesis, nature cannot be reduced to fundamental entities, like fundamental building blocks of matter, but has to be understood entirely through self-consistency. Things exist by virtue of their mutually consistent relationships, and all of physics has to follow uniquely from the requirement that its components be consistent with one another and with themselves. /51

> The mathematical framework of bootstrap physics is known as S-matrix theory. It is based on the concept of the S matrix, or 'scattering matrix,' which was originally proposed by Heisenberg in the 1940s and has been developed, over the

past two decades, into a complex mathematical structure, ideally suited to combine the principles of quantum mechanics with relativity theory. /51

This bootstrap philosophy not only abandons the idea of fundamental building blocks of matter, but accepts no fundamental entities whatsoever—no fundamental constants, laws, or equations. The material universe is seen as a dynamic web of interrelated events. None of the properties of any part of this web is fundamental; they all follow from the properties of the other parts, and the overall consistency of their interrelations determines the structure of the entire web. /51

My many interests beyond physics have kept me from doing research with Chew full time, and the University of California has never found it appropriate to support my part-time research, or to acknowledge my books and other publications as valuable contributions to the development and communication of scientific ideas. But I do not mind. Shortly after I returned to California, *The Tao of Physics* was published in the United States by Shambhala and then by Bantam Books, and has since become an international bestseller. The royalties from these editions and the fees for lectures and seminars, which I have given with increasing frequency, finally put an end to my financial difficulties, which had persisted through most of the seventies. /57

According to Chew, this bootstrapping will include the basic principles of quantum theory, our conception of macroscopic space-time, and eventually, even our conception of human consciousness. 'Carried to its logical extreme', writes Chew, 'the bootstrap conjecture implies that the existence of consciousness, along with all other aspects of nature, is necessary for self-consistency of the whole.' /61

Geoffrey Chew has had an enormous influence on my world view, my conception of science, and my way of doing research. Although I have repeatedly branched out very far from my original field of research, my mind is essentially a scientific mind, and my approach to the great variety of problems I have come to investigate has remained a scientific one, albeit within a very broad definition of science. It was Chew's influence, more than anything else, that helped me to develop such a scientific attitude in the most general sense of the term. /65

The following quote shows how Capra after the successful publication of *The Tao of Physics* begins to research for his next book, *The Turning Point* and how this book, that I personally find much better conceptually than *The Tao* was in the making from that time, and here is where we first read the name Gregory Bateson (1904-1980):

Over the years I experienced a profound change of perception and thought in this respect, and in the book that I finally wrote, *The Turning Point*, I no longer presented the new physics as a model for other sciences but rather as an important special case of a much more general framework, the framework of systems theory. /72

A central aspect of the emerging new paradigm, perhaps *the* central aspect, is the shift from objects to relationships. According to Bateson, relationship should be the basis of all definition; biological form is put together of relations and not of parts, and this is also how people think; in fact, he would say, it is the only way in which we can think. /78

The world gets much prettier as it gets more complicated, he [Bateson] would say. /79

One of Bateson's main aims in his study of epistemology was to point out that logic was unsuitable for the description of biological patterns. Logic can be used in very elegant ways to describe linear systems of cause and effect, but when causal sequences become circular, as they do in the living world, their description in terms of logic will generate paradoxes. This is true even for nonliving systems involving feedback mechanisms, and Bateson often used the thermostat as an illustration of his point. /80

Bateson would always insist that he was a monist, that he was developing a scientific description of the world which did not split the universe dualistically into mind and matter, or into any other separate entities. He often pointed out that Judeo-Christian religion, while boasting of monism, was essentially dualistic because it separated God from His creation. Similarly, he insisted that he had to exclude all other supernatural explanations because they would destroy the monistic structure of his science. /83

Bateson's most outstanding contributions to scientific thought, in my view, were his ideas about the nature of mind. He developed a radically new concept of mind, which represents for me the first successful attempt to really overcome the Cartesian split that has caused so many problems in Western thought and culture. /83

My first breakthrough in understanding Bateson's notion of mind came when I studied Ilya Prigogine's theory of self-organizing system. According to Prigogine, physicist, chemist, and Nobel laureate, the patterns of organization characteristic of living systems can be summarized in terms of a single dynamic principle, the principle of self-organization. /Id.

> A living organism is a self-organizing system, which means that its order is not imposed by the environment but is established by the system itself. In other words, self-organizing systems exhibit a certain degree of autonomy. This does not mean that they are isolated from their environment; on the contrary, they interact with it continually, but this interaction does not determine their organization; they are self-organizing. /84

As I mentioned at the start of this book review, another very good friend of Capra's was the British psychiatrist Ronald David Laing (1927-1989):

> This is where Laing parted company with most of his colleagues. He concentrated on the origins of mental illness by looking at the human condition—at the individual embedded in a network of multiple relationships—and thus addressed psychiatric problems in existential terms. Instead of treating schizophrenia and other forms of psychosis as diseases, he regarded them as special strategies that people invent in order to survive in unlivable situations. This view amounted to a radical change in perspective, which led Laing to see madness as a sane response to an insane social environment. In *The Politics of Experience* he articulated a trenchant social critique that resonated strongly with the critique of the counterculture and is as valid today as it was twenty years ago. /95

Finally, he broadly outlines his friendship and many conversations with Stanislav Grof. It is the absolutely best short description of Grof's astounding psychiatric research that I could find:

> Grof's cartography encompasses three major domains: the domain of 'psychodynamic' experiences, involving complex reliving of emotionally relevant memories from various periods of the individual's life; the domain of 'perinatal' experiences, related to the biological phenomena involved in the process of birth; and an entire spectrum of experiences going beyond individual boundaries and transcending the limitations of time and space, for which Grof has coined the term of 'transpersonal.' /100

> In perinatal experiences the sensations and feelings associated with the birth process may be relived in a direct, realistic way and may also emerge in the form of symbolic, visionary experiences. For example, the experience of enormous tensions that is characteristic of the struggle in the birth canal is often / accompanied by visions of titanic fights, natural disasters, and various images of destruction and self-destruction. To facilitate an understanding of the great complexity of physical symptoms, imagery, and experiential patterns, Grof has grouped them into four clusters, called perinatal matrices, which correspond to consecutive stages of the birth process. / 101

> 'A frequent error of current psychiatric practice,' Grof concluded, 'is to diagnose people as psychotics on the basis of the content of their experiences. My observations have convinced me that the idea of what is normal and what is pathological should not be based on the content and nature of people's experiences but on the way in which they are handled and on the degree to which a person is able to integrate these unusual experiences into his or her life.

Harmonious integration of transpersonal experiences is crucial to mental health, and sympathetic support and assistance in this process if of critical importance to a successful therapy. /122

'Once the therapeutic process has been initiated,' Grof went on, 'the role of the therapist is to facilitate the emerging experiences and help the client overcome resistances.'/ 122

Quotes

- Heisenberg, one of the founders of quantum theory and, along with Albert Einstein and Niels Bohr, one of the giants of modern physics, describes and analyzes the unique dilemma encountered by physicists during the first three decades of the century, when they explored the structure of atoms and the nature of subatomic phenomena. This exploration brought / them in contact with a strange and unexpected reality that shattered the foundations of their world view and forced them to think in entirely new ways. The material world they observed no longer appeared as a machine, made up of a multitude of separate objects, but rather as an indivisible whole; a network of relationships that included the human observer in an essential way. In their struggle to grasp the nature of atomic phenomena, scientists became painfully aware that their basic concepts, their language, and their whole way of thinking were inadequate to describe this new reality. /17-18

- Niels Bohr, sixteen years older than Heisenberg, was a man with supreme intuition and a deep appreciation for the mysteries of the world; a man influenced by the religious philosophy of Kierkegaard and the mystical writings of William James. He was never fond of axiomatic systems and declared repeatedly: 'Everything I say must be understood not as an affirmation but as a question.' Werner Heisenberg, on the other hand, had a clear, analytic, and mathematical mind and was / rooted philosophically in Greek thought, with which he had been familiar since his early youth. Bohr and Heisenberg represented complementary poles of the human mind, whose dynamic and often dramatic interplay was a unique process in the history of modern science and led to one of its greatest triumphs. /18-19

- Heisenberg himself played a decisive role in this development. He saw that the paradoxes in atomic physics appeared whenever one tried to describe atomic phenomena in classical terms, and he was bold enough to throw away the classical framework. In 1925 he published a paper in which he abandoned the conventional description of electrons within an atom in terms of their positions and velocities, which was used by Bohr and everybody else, and replaced it with a much more abstract framework, in which physical quantities were represented by mathematical structures called matrices. Heisenberg's 'matrix mechanics' was the first logically consistent formulation of quantum theory. It was supplemented one year later by a different formalism, worked out by Erwin Schrödinger and known as 'wave mechanics.' Both formalisms are logically consistent and are mathematically equivalent—the same atomic phenomenon can be described in two mathematically different languages. /19

- The term 'paradigm' from the Greek *paradeigma* ('pattern'), was used by Kuhn to denote a conceptual framework shared by a community of scientists and providing them with model problems and solutions. Over the next twenty years it would become very popular to speak of paradigms and paradigm shifts outside of science as well, and in *The Turning Point* I would use these terms in a very broad sense. A paradigm, for me, would mean the totality of thoughts, perceptions, and values that form a particular vision of reality, a vision that is the basis of a way of society organizing itself. /22

- The hippies opposed many cultural traits that we, too, found highly unattractive. To distinguish themselves from the crew cuts and polyester suits of the straight business executives they wore long hair, colorful and individualistic clothes, flowers, beads, and other jewelry. They lived naturally without disinfectants or deodorants, many of them vegetarians, many practicing yoga or some other form meditation. They would often bake their own bread or practice some craft. They were called 'dirty hippies' by the straight society but referred to themselves as 'the beautiful people.' Dissatisfied with a system of education that was designed to prepare young people for a society they rejected, many hippies dropped out of the educational system even though they were often highly talented. This subculture was immediately identifiable and tightly bound together. It had its own rituals, its music, poetry, and literature, a common fascination with spirituality and the occult, and the shared vision of a peaceful and beautiful society. Rock music and psychedelic drugs were powerful bonds that strongly influenced the art and life-style of the hippie culture. /23

▸ The sixties brought me without doubt the deepest and most radical personal experiences of my life: the rejection of conventional, 'straight' values; the closeness, peacefulness, and trust of the hippie community; the freedom of communal nudity; the expansion of consciousness through psychedelics and meditation; the playfulness and attention to the 'here and now'—all of which resulted in a continual sense of magic, awe, and wonder that, for me, will forever be associated with the sixties. /24

▸ The sixties were also the time when my political consciousness was raised. This happened first in Paris, where many graduate students and young research fellows were also active in the student movement that culminated in the memorable revolt that is still known simply as 'May 68.' I remember long discussions at the Science Faculty at Orsay, during which the students not only analyzed the Vietnam War and the Arab-Israeli war of 1967, but also questioned the power structure within the university and discussed alternative, nonhierarchical structures. /24

▸ I remember that my sympathy with the Black Power movement was aroused by a dramatic and unforgettable event shorty after we moved to Santa Cruz. We read in the newspaper that an unarmed black teenager had been brutally shot to death by a white policeman in a small record store in San Francisco. Outraged, my wife and I drove to San Francisco and went to the boy's funeral, expecting to find a large crowd of like-minded white people. Indeed, there was a large crowd, but to our great shock we found that, together with another two or three, we were the only whites. The congregation hall was lined with fierce-looking Black Panthers clad in black leather, arms crossed. The atmosphere was tense and we felt insecure and frightened. But when I approached one of the guards and asked whether it would be all right for us to attend the funeral, he looked straight into my eyes and said simply, 'You're welcome, brother, you're welcome!' /25

- After moving to California, I soon found out that Alan Watts was one of the heroes of the counterculture, whose books were on the shelves of most hippie communes, along with those of Carlos Castaneda, J. Krishnamurti, and Hermann Hesse. Although I had read books about Eastern philosophy and religion before reading Watts, it was he who helped me most to understand its essence. His books would take me as far as one could go with books and would stimulate me to go further through direct, nonverbal experience. /26

- The impact of Krishnamurti's physical appearance and charisma was enhanced and deepened by what he said. Krishnamurti was a very original thinker who rejected all spiritual authority and traditions. His teachings were quite close to those of Buddhism, but he never used any terms from Buddhism or from any other branch of traditional Eastern thought. The task he set himself was extremely difficult—to use language and reasoning in order to lead his audience beyond language and reasoning—and the way in which he went about it was highly impressive. /28

- The problem that Krishnamurti had solved for me, Zen-like with one stroke, is the problem most physicists face when confronted with the ideas of mystical traditions—how can one transcend thinking without losing one's commitment to science? It is the reason, I believe, that so many of my colleagues feel threatened by my comparisons between physics and mysticism. /31

- The Zen tradition, in particular, developed a system of nonverbal instruction through seemingly nonsensical riddles, called koans, which cannot be solved by thinking. They are designed precisely to stop the thought process and thus make the student ready for the nonverbal experience of reality. /32

- When I first read about the koan method in Zen training, it had a strangely familiar ring to me. I had spent many years studying another kind of paradox that seemed to play a similar role in the training of physicists. There were differences, of course. My own training as a physicist certainly had not had the intensity of Zen training. But then I thought about Heisenberg's account of the way in which physicists in the 1920s experienced the quantum paradoxes, struggling for understanding in a situation where nature alone was the teacher. The parallel was obvious and fascinating and, later on, when I learned more about Zen Buddhism, I found that it was indeed very significant. As in Zen, the solutions to the physicist's problems were hidden in paradoxes that could not be solved by logical reasoning but had to be understood in terms of a new awareness, the awareness of the atomic reality. Nature was their teacher and, like the Zen masters, she did not provide any statements; she just provided the riddles. / 32

▸ Later on, I also came to understand why quantum physics and Eastern mystics were faced with similar problems and went through similar experiences. Whenever the essential nature of things is analyzed by the intellect it will seem absurd or paradoxical. This has always been recognized by mystics but has become a problem in science only very recently. For centuries, the phenomena studied in science belonged to the scientists' everyday environment and thus to the realm of their sensory experience. Since the images and concepts of their language were abstracted from this very experience, they were sufficient and adequate to describe the natural phenomena. / 33

▸ In the twentieth century, however, physicists penetrated deep into the submicroscopic world, into realms of nature far removed from our macroscopic environment. Our knowledge of matter at this level is no longer derived from direct sensory experience, and therefore our ordinary language is no longer adequate to describe the observed phenomena. Atomic physics provided the scientists with the first glimpses of the essential nature of things. Like the mystics, physicists were now dealing with a nonsensory experience of reality and, like the mystics, they had to face the paradoxical aspects of this experience. From then on, the models and images of modern physics became akin to those of Western philosophy. / 33

▸ Shortly before leaving California I had designed a photo-montage—a dancing Shiva superimposed on tracks of colliding particles in a bubble chamber—to illustrate my experience of the cosmic dance on the beach. One day I sat in my tiny room near Imperial College and looked at this beautiful picture, and suddenly I had a very clear realization. I knew with absolute certainty that the parallels between physics and mysticism, which I had just begun to discover, would someday be common knowledge; I also knew that I was best placed to explore these parallels

thoroughly and to write a book about them. I resolved there
and then to write that book, but I also decided that I was not
yet ready to do so. I would first study my subject further and
write a few articles about it before attempting the book. / 35

- The Taoist sages concentrated their attention fully on the
  observation of nature in order to discern the 'characteristics
  of the Tao.' In doing so they developed an attitude that was
  essentially scientific; only their deep mistrust of the analytic
  method of reasoning prevented them from constructing
  proper scientific theories. Nevertheless, their careful
  observation of nature, combined with a strong mystical
  intuition, led them to profound insights which are confirmed
  by modern scientific theories. The deep ecological wisdom,
  the empirical approach, and the special flavor of Taoism,
  which I can best describe as 'quiet ecstasy,' were enormously
  attractive to me, and so Taoism quite naturally became the
  way for me to follow. / 36

- Castaneda, too, exerted a strong influence on me in those
  years, and his books showed me yet another approach to the
  spiritual teachings of the East. I found the teachings of the
  American Indian traditions, expressed by the legendary
  Yaqui sage Don Juan, very close to those of the Taoist
  tradition, transmitted by the legendary sages Lao Tzu and
  Chuang Tzu. The awareness of being embedded in the
  natural flow of things and the skill to act accordingly are
  central to both traditions. As the / Taoist sage flows in the
  current of the Tao, the Yaqui 'man of knowledge' needs to be
  light and fluid to 'see' the essential nature of things. / 36-37

- The strongest influence of Buddhist tradition on my own
  thinking has been the emphasis on the central role of
  compassion in the attainment of knowledge. According to
  the Buddhist view, there can be no wisdom without
  compassion, which means for me that science is of no value
  unless it is accompanied by social concern. / 37

- Although the years 1971 and 1972 were very difficult for me,
  they also were very exciting. I continued my life as part-time
  physicist and part-time hippie, doing research in particle
  physics at Imperial College while also pursuing my larger
  research in an organized and systematic way. I managed to
  get several part-time jobs—teaching high-energy physics to a
  group of engineers, translating technical texts from English
  into German, teaching mathematics to high school girls—

> which made enough money for me to survive but did not allow for any material luxury. My life during those two years was very much like that of a pilgrim; its luxuries and joys were not those of the material plane. What carried me through this period was a strong belief in my vision and a conviction that my persistence would eventually be rewarded. During those two years I always had a quote from the Taoist sage Chuang Tzu pinned to my wall: 'I have sought a ruler who would employ me for a long time. That I have not found one shows the character of the time.' /37

▸ I packed my suit, shirts, leather shoes, and physics papers in a bag, put on my patched jeans, sandals, and flowered shirt, and hit the road. The weather was superb and I greatly enjoyed traveling through Europe the slow way, meeting lots of people and visiting beautiful old towns on the way. My overriding experience on this trip, the first to Europe after two years of California, was the realization that European national borders are rather artificial divisions. I noticed that the language, customs, and physical characteristics of the people did not change abruptly at the borders, but rather gradually, and that the people on either side of the border often had much more in common with each other than, say, with the inhabitants of the capitals of their countries. /38

▸ Today *The Tao of Physics* is an international bestseller and is often praised as a classic that has influenced many other writers. But when I planned to write it, it was extremely difficult for me to find a publisher. Friends in London who were writers suggested that I should first look for a literary agent, and even that took considerable time. When I finally found an agent who agreed to take on this unusual project, he told me that he would need an outline of the book plus three sample chapters to offer to prospective publishers. This put me in a great dilemma. /45

▸ In the meantime, my agent offered the manuscript to the major publishers in London and New York, all of / whom turned it down. After a dozen rejections, a small but enterprising London publishing firm, Wildwood House, accepted the proposal and paid me an advance that gave me sufficient support to write the entire book. /45-46

▸ According to the bootstrap hypothesis, nature cannot be reduced to fundamental entities, like fundamental building blocks of matter, but has to be understood entirely through self-consistency. Things exist by virtue of their mutually consistent relationships, and all of physics has to follow uniquely from the requirement that its components be consistent with one another and with themselves. /51

> The mathematical framework of bootstrap physics is known as S-matrix theory. It is based on the concept of the S matrix, or 'scattering matrix,' which was originally proposed by Heisenberg in the 1940s and has been developed, over the past two decades, into a complex mathematical structure, ideally suited to combine the principles of quantum mechanics with relativity theory. /51

> This bootstrap philosophy not only abandons the idea of fundamental building blocks of matter, but accepts no fundamental entities whatsoever—no fundamental constants, laws, or equations. The material universe is seen as a dynamic web of interrelated events. None of the properties of any part of this web is fundamental; they all follow from the properties of the other parts, and the overall consistency of their interrelations determines the structure of the entire web. /51

> My many interests beyond physics have kept me from doing research with Chew full time, and the University of California has never found it appropriate to support my part-time research, or to acknowledge my books and other publications as valuable contributions to the development and communication of scientific ideas. But I do not mind. Shortly after I returned to California, *The Tao of Physics* was published in the United States by Shambhala and then by Bantam Books, and has since become an international bestseller. The royalties from these editions and the fees for lectures and seminars, which I have given with increasing frequency, finally put an end to my financial difficulties, which had persisted through most of the seventies. /57

> According to Chew, this bootstrapping will include the basic principles of quantum theory, our conception of macroscopic space-time, and eventually, even our conception of human consciousness. 'Carried to its logical extreme,' writes Chew, 'the bootstrap conjecture implies that the existence of consciousness, along with all other aspects of nature, is necessary for self-consistency of the whole.' /61

- Quantum mechanics has something intrinsically discrete about it, whereas the idea of space-time is continuous. I believe that if you try to state the principles of quantum mechanics after having accepted space-time as an absolute truth, then you will get into difficulties. /62

- Bohm's starting point is the notion of 'unbroken wholeness,' and his aim is to explore the order he believes to be inherent in the cosmic web of relations at a deeper, 'nonmanifest' level. He calls this order 'implicate,' or 'enfolded,' and describes it with the analogy of a hologram, in which each part, in some sense, contains the whole. If any part of a hologram is illuminated, the entire image will be reconstructed, although it will show less detail than the image obtained from the complete hologram. In Bohm's view the real world is structured according to the same general principles, with the whole enfolded in each of its parts. /64

- Bohm's theory is still tentative, but there seems to be an intriguing kinship, even at the preliminary stage, between his theory and the implicate order and Chew's bootstrap theory. Both approaches are based on a view of the world as a dynamic web of relations; both attribute a central role to the notion of order; both use matrices to represent change and transformation, and topology to classify categories of order. /64

- Geoffrey Chew has had an enormous influence on my world view, my conception of science, and my way of doing research. Although I have repeatedly branched out very far from my original field of research, my mind is essentially a scientific mind, and my approach to the great variety of problems I have come to investigate has remained a scientific one, albeit within a very broad definition of science. It was Chew's influence, more than anything else, that helped me to develop such a scientific attitude in the most general sense of the term /65

- It appears that the science of the future will no longer need any firm foundations, that the metaphor of the building will be replaced by that of the web, or network, in which no part is more fundamental than any other part. Chew's bootstrap theory is the first scientific theory in which such a 'web philosophy' has been formulated explicitly, and he agreed in a recent conversation that abandoning the need for firm foundations may be a major shift and deepest change in natural science. /66

- A methodology that does not use well-defined questions and recognizes no firm foundation of one's knowledge does indeed seem highly unscientific. What turns it into a scientific endeavor is another essential element of Chew's approach, which represents another major lesson I learned from him—recognition of the crucial role of approximation in scientific theories. /67

- Over the years I experienced a profound change of perception and thought in this respect, and in the book that I finally wrote, *The Turning Point*, I no longer presented the new physics as a model for other sciences but rather as an important special case of a much more general framework, the framework of systems theory. /72

- A central aspect of the emerging new paradigm, perhaps *the* central aspect, is the shift from objects to relationships. According to Bateson, relationship should be the basis of all definition; biological form is put together of relations and not of parts, and this is also how people think; in fact, he would say, it is the only way in which we can think. /78

- The world gets much prettier as it gets more complicated, he [Bateson] would say. /79

- One of Bateson's main aims in his study of epistemology was to point out that logic was unsuitable for the description of biological patterns. Logic can be used in very elegant ways to describe linear systems of cause and effect, but when causal sequences become circular, as they do in the living world, their description in terms of logic will generate paradoxes. This is true even for nonliving systems involving feedback mechanisms, and Bateson often used the thermostat as an illustration of his point. /80

- Bateson would always insist that he was a monist, that he was developing a scientific description of the world which did not split the universe dualistically into mind and matter, or into any other separate entities. He often pointed out that Judeo-Christian religion, while boasting of monism, was essentially dualistic because it separated God from His creation. Similarly, he insisted that he had to exclude all

- other supernatural explanations because they would destroy the monistic structure of his science. / 83

- Bateson's most outstanding contributions to scientific thought, in my view, were his ideas about the nature of mind. He developed a radically new concept of mind, which represents for me the first successful attempt to really overcome the Cartesian split that has caused so many problems in Western thought and culture. / 83

- My first breakthrough in understanding Bateson's notion of mind came when I studied Ilya Prigogine's theory of self-organizing systems. According to Prigogine, physicist, chemist, and Nobel laureate, the patterns of organization characteristic of living systems can be summarized in terms of a single dynamic principle, the principle of self-organization. A living organism is a self-organizing system, which means that its order is not imposed by the environment but is established by the system itself. In other words, self-organizing systems exhibit a certain degree of autonomy. This does not mean that they are isolated from their environment; on the contrary, they interact with it continually, but this interaction does not determine their organization; they are self-organizing. / 84

- Over the last fifteen years, a theory of self-organizing systems has been developed in considerable detail by a number of researchers from various disciplines under the leadership of Prigogine. My understanding of this theory was helped enormously by extensive discussion with Erich Jantsch, a system theorist who was one of Prigogine's principle disciples and interpreters. / 84

- It was Erich Jantsch who pointed out to me the connection between Prigogine's concept of self-organization and Bateson's concept of mind. Indeed, when I compared Prigogine's criteria for self-organizing systems to Bateson's criteria of mental process, I found that the two sets of criteria were very similar; in fact, they seemed close to being identical. I realized immediately that this meant that mind and self-organization were merely different aspects of one and the same phenomenon, the phenomenon of life. / 85

- From that moment on my understanding of the relationship between mind and life, or mind and nature, as Bateson put it, continued to deepen, and with it came an increased appreciation of the richness and beauty of Bateson's thought. I realized fully why it is impossible for him to separate mind and matter. When Bateson looked at the living world, he saw its principles of organization as being essentially mental, with mind being immanent in matter at all levels of life. He thus arrived at the unique synthesis of notions of mind with notions of matter; a synthesis that was, as he liked to point out, neither mechanical nor supernatural. /85

- This is where Laing parted company with most of his colleagues. He concentrated on the origins of mental illness by looking at the human condition—at the individual embedded in a network of multiple relationships—and thus addressed psychiatric problems in existential terms. Instead of treating schizophrenia and other forms of psychosis as diseases, he regarded them as special strategies that people invent in order to survive in unlivable situations. This view amounted to a radical change in perspective, which led Laing to see madness as a sane response to an insane social environment. In *The Politics of Experience* he articulated a trenchant social critique that resonated strongly with the critique of the counterculture and is as valid today as it was twenty years ago. /95

- The absence of any distinct drug-specific effects and the enormous range of phenomena that occur during these sessions have convinced me that LSD is best understood as a powerful unspecific amplifier, or catalyst, of mental processes, which facilitates the emergence of unconscious material from different levels of the human psyche. /97

- Grof's cartography encompasses three major domains: the domain of 'psychodynamic' experiences, involving complex reliving of emotionally relevant memories from various periods of the individual's life; the domain of 'perinatal' experiences, related to the biological phenomena involved in the process of birth; and an entire spectrum of experiences going beyond individual boundaries and transcending the limitations of time and space, for which Grof has coined the term of 'transpersonal.' /100

- In perinatal experiences the sensations and feelings associated with the birth process may be relived in a direct, realistic way and may also emerge in the form of symbolic, visionary experiences. For example, the experience of enormous tensions that is characteristic of the struggle in the birth canal is often / accompanied by visions of titanic fights, natural disasters, and various images of destruction and self-destruction. To facilitate an understanding of the great complexity of physical symptoms, imagery, and experiential patterns, Grof has grouped them into four clusters, called perinatal matrices, which correspond to consecutive stages of the birth process. / 101

- 'A frequent error of current psychiatric practice,' Grof concluded, 'is to diagnose people as psychotics on the basis of the content of their experiences. My observations have convinced me that the idea of what is normal and what is pathological should not be based on the content and nature of people's experiences but on the way in which they are handled and on the degree to which a person is able to integrate these unusual experiences into his or her life. Harmonious integration of transpersonal experiences is crucial to mental health, and sympathetic support and assistance in this process is of critical importance to a successful therapy.' / 122

- 'Once the therapeutic process has been initiated,' Grof went on, 'the role of the therapist is to facilitate the emerging experiences and help the client overcome resistances.' / 122

- Laing then went on to speculate about a new kind of language that would be appropriate for the new science. He pointed out to me that conventional scientific language is descriptive, whereas language to share experience needs to be *depictive*. It would be a language more akin to poetry, or even to music, which would depict an experience directly, conveying, somehow, its qualitative character. / 139

- In addition to the *yin/yang* system,' Lock continued, 'the Chinese used a system called *Wu Hsing* to describe the great patterned order of the cosmos. This is usually translated as the 'five elements,' but Porkert has translated it as the 'five evolutive phases,' which conveys the Chinese idea of dynamic relationships much better. Lock explained that an intricate correspondence system was derived from the five phases, which extended to the entire universe. The seasons, the atmospheric influences, colors, sounds, parts of the body, emotional states, social relations, and numerous other phenomena were all classified into five types related to the five phases. When the five-phase theory was fused with the *yin/yang* cycles, the result was an elaborate system in which every aspect of the universe was described as a well-defined part of a dynamically patterned whole. This system, Lock explained, formed the theoretical foundation for the diagnosis and treatment of illness. /157

- 'So what is illness in the Chinese view?' I asked her. 'Illness is an imbalance which occurs when the *ch'i* does not circulate properly. This is another important concept in Chinese natural philosophy, as you know. The word means literally 'vapor' and was used in ancient China to describe the vital breath, or energy, animating the cosmos. The flow and fluctuation of *ch'i* keep a person alive, and there are definite pathways of *ch'i*, the well-known meridians, along which lie the acupuncture points.' /157

- Throughout my conversations with Margaret Lock I had the strong feeling that the philosophy underlying East Asian medicine is very much in agreement with the new paradigm that is now emerging from modern Western science. Moreover, it was evident to me that many of its characteristics should be important aspects of our new holistic medicine as well—for example, the view of health as a process of dynamic balance, the attention given to the continual interplay between the human organism and its natural environment, and the importance of preventive medicine. /169

- It was clear to me that in such a holistic approach to health and healing the concept of health itself would have to be much more subtle than in the biomedical model, where health is defined as the absence of disease and disease is seen as a malfunctioning of biological mechanisms. The holistic concept would picture health as reflecting the state of the whole organism, mind and body, and would also see it in relation to the organism's environment. I also realized that the new concept of health should be a dynamic concept, seeing health as a process of dynamic balance and acknowledging, somehow, the healing forces inherent in living organisms. /172

- The basic philosophy of the Simonton approach affirms that the development of cancer involves a number of interdependent psychological and biological processes, that these processes can be recognized and understood, and that the sequence of events which leads to illness can be reversed to lead the organism back to the healthy state. To do so, the Simontons help their patients to become aware of the wider context of their illness, identify the major stresses in their lives, and develop a positive attitude about the effectiveness of the treatment and the potency of the body's defenses. /175

▸ When I asked her to elaborate her point, Henderson contended that there is no single cause of inflation, but that several major sources can be identified, all of which involve variables that have been excluded from current economic models. The first source, she pointed out, has to do with the fact—still ignored by most economists—that wealth is based on natural resources and energy. As the resource base declines, raw materials and energy must be extracted from ever more degraded and inaccessible reservoirs, and thus more and more capital is needed for the extraction process. Consequently, the inevitable decline of natural resources is accompanied by an unremitting climb of the price of resources and energy, which becomes one of the main driving forces of inflation. /254

Chapter Eight
The Web of Life

The Web of Life

A New Scientific Understanding of Living Systems
New York: Anchor Books, 1997

Review

The Web of Life is the book in which Capra defined his approach to ecology, thereby making ecology, or *deep ecology,* a concept that is part of a new science paradigm, powerfully introduced and promoted by one of the most important science theorists of our times.

What is *deep ecology and why do we need it?* Capra writes:

> Whereas the old paradigm is based on anthropocentric (human-centered) values, deep ecology is grounded in ecocentric (earth-centered) values. It is a worldview that acknowledges the inherent value of nonhuman life./11

> Such a deep ecological ethics is urgently needed today, and especially in science, since most of what scientists do is not life-furthering and life-preserving but life-destroying. With physicists designing weapon systems that threaten to wipe out life on the planet, with chemists contaminating the global environment, with biologists releasing new and unknown types of microorganisms without knowing the consequences, with psychologists and other scientists torturing animals in the name of scientific progress—with all these activities going on, it seems most urgent to introduce 'ecoethical' standards into science./Id.

This book's quest is enormous in that it requires modern science to fundamentally shift its regard upon nature, and upon living! Our regard upon nature has been conditioned by patriarchy since about five thousand years, and it's a rather defensive, distorted, schizophrenic, and reductionist regard. Capra looked back in history and found amazing early intuitions and truths propagated by our great thinkers, poets and philosophers, such as for example Immanuel Kant, Johann Wolfgang von Goethe or William Blake.

The understanding of organic form also played an important role in the philosophy of Immanuel Kant, who is often considered the greatest of the modern philosophers. An idealist, Kant separated the phenomenal world from a world of 'things-in-themselves.' He believed that science could offer only mechanical explanations, but he affirmed that in areas where such explanations were inadequate, scientific knowledge needed to be supplemented by considering nature as being purposeful./21

Capra wondered why our science and technologies are so deeply hostile toward our globe, which we call *Mother Earth* after all, and so little caring for its preservation? He found conclusive answers in ancient traditions that fostered what we call today a *Gaia* worldview, a respectful attitude toward the earth, the mother, the yin energy and generally, female values:

> The view of the Earth as being alive, of course, has a long tradition. Mythical images of the Earth Mother are among the oldest in human religious history. Gaia, the Earth Goddess, was revered as the supreme deity in early, pre-Hellenic Greece. Earlier still, from the Neolithic through the Bronze Ages, the societies of 'Old Europe' worshipped numerous female deities as incarnations of Mother Earth./22

This is how Capra, always grounded in common sense and meaningful retrospection, smoothly introduces the novice reader to the concept of *systems research* or the *systems view of life*.

Post-matriarchal thought, which was naturally systemic, can be traced from the *Atomistic Worldview*

(Democritus), over the *Cartesian Worldview* (Newton, La Mettrie, René Descartes) and *Relativistic Worldview* (Einstein, Planck, Heisenberg), to the *Systemic Worldview* (Bohm, Bateson, Grof, Capra, Laszlo, etc.) and the *Holistic Worldview* (Talbot, Goswami, McTaggart, etc.).

In all systems, we have to deal with different levels of complexity that are woven in each other, thus rendering it almost impossible to dissect parts of the system for closer research without disturbing the system. This means that, contrary to earlier vivisectionist science, we need to leave the system intact and focus our research onto the whole of it—which makes it all so complex, but this very complexity renders justice to nature!

As a result, we had to develop a new mathematics, which today is called the mathematics of complexity, in order to deal with the high complexity levels in living systems. This also means that our chief scientific method—deductive analysis—is inadequate for any inquiry in the functionality of living systems, because they are networks within networks and can only be grasped scientifically through understanding their *properties*.

> According to the systems view, the essential properties of an organism, or living system, are properties of the whole, which none of the parts have. They arise from the interactions and relationships among the parts. These properties are destroyed when the system is dissected, either physically or theoretically, into isolated elements. Although we can discern individual parts in any system, these parts are not isolated, and the nature of the whole is always

different from the mere sum of its parts. (...) The great shock of twentieth-century science has been that systems cannot be understood by analysis. The properties of the parts are not intrinsic properties but can be understood only within the context of the larger whole. Thus the relationship between the parts and the whole has been reversed./29

At each scale, under closer scrutiny, the nodes of the network reveal themselves as smaller networks. We tend to arrange these systems, all nesting within larger systems in a hierarchical scheme by placing the larger systems above the smaller ones in pyramid fashion. But this is a human projection. In nature there is no 'above' or 'below,' and there are no hierarchies. There are only networks nesting within other networks./35

This means that living systems are not, as most of our governmental and societal organization, hierarchical, but network-based, and thus structured not vertically but horizontally by 'neuronally' linking segments to larger molecular structures that distribute information instantly over the whole of the network. You can also say that a living network is a system of 'total information sharing' where there is not one single molecule that is uninformed at any point in time and space.

The fact that horizontal networks are nested within other horizontal networks, while the different networks all possess a different level of complexity, makes research so intricate. This is inter alia why high-performance computers have greatly aided in developing systems theory. But the most revolutionary insight here is that our usual habit of dissecting parts from a whole for further scrutiny and scientific investigation does not work with living systems. Why is this so?

> Ultimately—as quantum physics showed so dramatically—there are no parts at all. What we call a part is merely a pattern in an inseparable web of relationships. Therefore the shift from the parts to the whole can also be seen as a shift from objects to relationships./37

Hence, the whole of our approach to scientific investigation has to shift from an object-based to a relationship-based research approach when we deal with *living systems*. This requires researchers to change their inner setup which is exactly what quantum physics revealed to us, that is, the observer's belief system will be reflected in the outcome of the research.

And there is one more crucial element in systems research that Capra explains and elucidates. It is what we already learnt within the revolutionary reframing of science by quantum physics, the fact namely that in approaching quantum reality, and organic behavior, we have to learn the mathematics of probability. What is probability? It is the approximation of behavior. Dealing with approximations means that we leave the certainty principle and venture into what Heisenberg called the *uncertainty principle*. Giving up certainty triggers fear. This fear was very vividly described by Max Planck and Werner Heisenberg when the paradigm began to shift and quantum physics slowly but definitely began to undermine traditional physics. When we abandon certainty, we begin to grasp the notion of *approximation*, and of probability, and accordingly, we will shift our mathematical constructs when we deal with open systems.

> What makes it possible to turn the systems approach into a science is the discovery that there is approximate knowledge. This insight is crucial to all of modern science. The old paradigm is based on the Cartesian belief in the certainty of scientific knowledge. In the new paradigm it is recognized that all scientific concepts and theories are limited and approximate. Science can never provide any complete and definite understanding./41
>
> Unlike closed systems, which settle into a state of thermal equilibrium, open systems maintain themselves far from equilibrium in this 'steady state' characterized by continual flow and change./48

Living systems are open systems, which means that their main characteristic is *change and flow*, and not continuity and static behavior. And they are far from equilibrium, which is the single most revolutionary discovery of systems research. It means living systems are constantly struggling against decay, and decay means equilibrium. When we extrapolate this insight from organic systems into our metaphysical reality, we see that it applies also to human beings, and even to religions. When we are settled and satiated, we are not alive. This is what it all boils down to. So this profound insight from systems research may help us to survive in a state far from equilibrium, putting our assuredness or false assuredness away, to stay with a beginner's mind, as it's so wistfully expressed in Zen. Our universe is a basically *patterned universe*, so is human intelligence.

But what are patterns? Capra explains the importance of pattern when he explores the meaning of *self-organization*, which is one major characteristic of living systems. In order to scientifically explain pattern we need to change or for the least upgrade our basic toolset of scientific investigation. Capra explains:

> To understand the phenomenon of self-organization, we first need to understand the importance of pattern. The idea of a pattern of organization—a configuration of relationships characteristic of a particular system—became the explicit focus of systems thinking in cybernetics and has been a crucial concept ever since. From the systems point of view, the understanding of life begins with the understanding of pattern./80

> In the study of structure we measure and weigh things. Patterns, however, cannot be measured or weighed; they must be mapped. To understand a pattern we must map a configuration of relationships. In other words, structure involves quantities, while pattern involves qualities./81

The systems view of life really involves a radical change in our scientific thinking because traditional science was quantity-based and measure-oriented, while systemic science is quality-based and relationship-oriented.

Capra exemplifies this truth by looking at the properties involved in the scientific focus of both static and systemic science theory. In this context, we should look at feedback loops as an important self-regulatory function in living systems. This is important because without feedback loops, living systems could not be self-organizing. Capra explains:

> Systemic properties are properties of pattern. What is destroyed when a living organism is dissected is its pattern. The components are still there, but the configuration of relationships among them— the pattern—is destroyed, and thus the organism dies./81

> Because networks of communication may generate feedback loops, they may acquire the ability to regulate themselves. For example, a community that maintains an active network of communication will learn from its mistakes, because the consequences of a mistake will spread through the network and return to the / source along feedback loops. Thus the community can correct its mistakes, regulate itself, and organize itself. Indeed, self-organization has emerged as perhaps the central concept in the systems view of life, and like the concepts of feedback and self-regulation, it is linked closely to networks. The pattern of life, we might say, is a network pattern capable of self-organization. This is a simple definition, yet it is based on recent discoveries at the very forefront of science./82-83

Another central point in this book is Capra's focus upon the intrinsic quality of living systems as *nonlinear systems* that require, to be understood, an equally *nonlinear* mathematical approach. One early realization of mathematical nonlinearity was the introduction of the fractal in mathematics. In fact, in my exchanges with the Swiss mathematician Peter Meyer who was the collaborator of Terence McKenna for the realization of the *Timewave Zero* calculus as a part of *Novelty Theory*, I learnt that time is a fractal. Capra explains:

> The great fascination exerted by chaos theory and fractal geometry on people in all disciplines—from scientists to managers to artists—may indeed be a hopeful sign that the isolation of mathematics is ending. Today the new mathematics of complexity is / making more and more people realize that mathematics is much more than dry formulas; that the understanding of pattern is crucial to understand the living world around us; and that all questions of pattern, order, and complexity are essentially mathematical./152-153

After having elucidated that systems research involves a process-based scientific approach rather than an object-based one, Capra presents the perhaps most important research topic in this book: the reinvestigation of *cognition* based on the insights from systems research. Capra pursues:

> The identification of mind, or cognition, with the process of life is a radically new idea in science, but it is also one of the deepest and most archaic intuitions of humanity. In ancient times the rational human mind was seen as merely one aspect of the immaterial soul, or spirit./264

In fact, the whole debate about information processing, vividly criticized in the early writings of think tank

Edward de Bono, and the even larger debate about cybernetics make it all clear that cognition is currently in a process of reevaluation:

> The computer model of cognition was finally subjected to serious questioning in the 1970's when the concept of self-organization emerged. (...) These observations suggested a shift of focus—from symbols to connectivity, from local rules to global coherence, from information processing to the emergent properties of neural networks./266

In my scientific exploration of emotions, I have revisited our scientific grasp of emotions, as it was coined within a fragmented and reductionist science paradigm. Fritjof Capra comprehensively explains that emotions are not singular elements but coherently organized within a *patterned system* in which cognition and response are intertwined in a self-regulatory and organic whole:

> The range of interactions a living system can have with its environment defines its 'cognitive domain.' Emotions are an integral part of this domain. For example, when we respond to an insult by getting angry, that entire pattern of physiological processes—a red face, faster breathing, trembling, and so on—is part of cognition. In fact, recent research strongly indicates that there is an emotional coloring to every cognitive act./269

The most important fact that systems theory teaches us about cognition is that it does not work like a computer processes information. Information processing, already declared years ago 'an obsession of modern science' by

Edward de Bono, is quite a misnomer because our brain does not 'process' information as a computer does.

> A computer processes information, which means that it manipulates symbols based on certain rules. The symbols are distinct elements fed into the computer from outside, and during the information processing there is no change in the structure of the machine. The physical structure of the computer is fixed, determined by its design and construction. The nervous system of a living organism ... interacts with its environment by continually modulating its structure, so that at any moment its physical / structure is a record of previous structural changes. The nervous system does not process information from the outside world but, on the contrary, brings forth a world in the process of cognition. / 274-275

Capra then answers to the debate about *artificial intelligence* and the myths it creates in the minds of masses of people:

> A lot of confusion is caused by the fact that computer scientists use words such as intelligence, memory, and language to describe computers, thus implying that these expressions refer to the human phenomena we know well from experience. This is a serious misunderstanding. For example, the very essence of intelligence is to act appropriately when a problem is not clearly defined and solutions are not evident. Intelligent human behavior in such situations is based on common sense, accumulated from lived experience. Common sense, however, is not available to computers because of their blindness of abstraction and the intrinsic limitations / of formal operations, and therefore

it is impossible to program computers to be intelligent./275-276

Real intelligence is human, and original, not mechanical and artificial! True intelligence is contextual, as language is. No computer can understand meaning. A rat's intelligence is a million times closer to that of man than that of the most powerful and sophisticated computer. Capra notes:

> The reason is that language is embedded in a web of social and cultural conventions that provides an unspoken context of meaning. We understand this context because it is common sense to us, but a computer cannot be programmed with common sense and therefore does not understand language./276

> Mind is not a thing but a process—the process of cognition, which is identified with the process of life. The brain is a specific structure through which this process operates. Thus the relationship between mind and brain is one between process and structure./278

Now, let us look at what sustainability means in systems research. A system is sustainable when it's not only functional but also well integrated in a greater continuum so that it has a good prognosis for survival, for continuity. Capra writes:

> Partnership is an essential characteristic of sustainable communities. The cyclical exchanges of energy and resources in an ecosystem are sustained by pervasive cooperation. Indeed, we have seen that since the creation of

the first nucleated cells over two billion years ago, life on Earth has proceeded through ever more intricate arrangements of cooperation and coevolution. Partnership—the tendency to associate, establish links, live inside one another, and cooperate—is one of the hallmarks of life./301

Partnership and cooperation were indeed alien words under patriarchy but before that time, they were imbedded in pre-patriarchal cultures, such as the *Minoan Civilization*, and thus what we get today is a return to the sources.

Quotes

- There are solutions to the major problems of our time, some of them even simple. But they require a radical shift in our perceptions, our thinking, our values. And, indeed, we are now at the beginning of such a fundamental change of worldview in science and society, a change of paradigms as radical as the Copernican revolution. But this realization has not yet dawned on most of our political leaders. The recognition that a profound change of perception and thinking is needed if we are to survive has not yet reached most of our corporate leaders, either, or the administrators and professors of our large universities. /4

- Whereas the old paradigm is based on anthropocentric (human-centered) values, deep ecology is grounded in ecocentric (earth-centered) values. It is a worldview that acknowledges the inherent value of nonhuman life. /11

- Such a deep ecological ethics is urgently needed today, and especially in science, since most of what scientists do is not life-furthering and life-preserving but life-destroying. With physicists designing weapon systems that threaten to wipe out life on the planet, with chemists contaminating the global environment, with biologists releasing new and unknown types of microorganisms without knowing the consequences, with psychologists and other scientists torturing animals in the name of scientific progress—with all these activities going on, it seems most urgent to introduce 'ecoethical' standards into science. /11

- The emphasis on the parts has been called mechanistic, reductionist, or atomistic; the emphasis on the whole holistic, organismic, or ecological. In twentieth-century science the holistic perspective has become known as 'systemic' and the way of thinking it implies as 'systems thinking.' /17

- At the dawn of Western philosophy and science, the Pythagoreans distinguished 'number,' or pattern, from substance, or matter, viewing it as something that limits matter and gives it shape. As Gregory Bateson put it:
  —The argument took the shape of 'Do you ask what it's made of—earth fire, water, etc.?' Or do you ask, 'What is its pattern?' Pythagoreans stood for inquiring into pattern rather than inquiring into substance. /18

- William Blake, the great mystical poet and painter who exerted a strong influence on English Romanticism, was a passionate critic of Newton. He summarized his critique in these celebrated lines:
  —May God us keep from single vision and Newton's sleep. /21

- 'Each creature,' wrote Goethe, 'is but a patterned gradation (Schattierung) of one great harmonious whole.' /21

- The Romantic artists were concerned mainly with a qualitative understanding of patterns, and therefore they placed great emphasis on explaining the basic properties of life in terms of visualized forms. Goethe, in particular, felt that visual perception was the door to understanding organic form. /21

- The understanding of organic form also played an important role in the philosophy of Immanuel Kant, who is often considered the greatest of the modern philosophers. An idealist, Kant separated the phenomenal world from a world of 'things-in-themselves.' He believed that science could offer only mechanical explanations, but he affirmed that in areas where such explanations were inadequate, scientific knowledge needed to be supplemented by considering nature as being purposeful. /21

- The view of the Earth as being alive, of course, has a long tradition. Mythical images of the Earth Mother are among the oldest in human religious history. Gaia, the Earth Goddess, was revered as the supreme deity in early, pre-Hellenic Greece. Earlier still, from the Neolithic through the Bronze Ages, the societies of 'Old Europe' worshipped numerous female deities as incarnations of Mother Earth. /22

▸ The German embryologist Hans Driesch initiated the opposition to mechanistic biology at the turn of the century with his pioneering experiments on sea urchin eggs, which led him to formulate the first theory of vitalism. When Driesch destroyed one of the cells of an embryo at the very early two-celled stage, the remaining cell developed not into half a sea urchin, but into a complete but smaller organism. Similarly, complete smaller organisms developed after the destruction of two or three cells in four-celled embryos. Driesch realized that his sea urchins eggs had done what a machine could never do: they had regenerated wholes from some of their parts. /26

▸ This is, in fact, the root meaning of the word 'system,' which derives from the Greek *synhistanai* ('to place together'). To understand things systemically literally means to put them into a context, to establish the nature of their relationships. /27

▸ Indeed, an outstanding property of all life is its tendency to form multileveled structures of systems within systems. Each of these forms a whole with respect to its parts while at the same time being a part of a larger whole. Thus cells combine to form tissues, tissues to form organs, and organs to form organisms. These in turn exist within social systems and ecosystems. Throughout the living world we find living systems nesting within other living systems. /28

▸ What the early systems thinkers recognized very clearly is the existence of different levels of complexity with different kinds of laws operating at each level. Indeed, the concept of 'organized complexity' became the very subject of the systems approach. At each level of complexity the observed phenomena exhibit properties that do not exist at the lower level. For example, the concept of temperature, which is central to thermodynamics, is meaningless at the level of individual atoms, where the laws of quantum theory operate. Similarly, the taste of sugar is not present in the carbon, hydrogen, and oxygen atoms that constitute its components. In the early 1920's the philosopher C.D. Broad coined the / term 'emergent properties' for those properties that emerge at a certain level of complexity but do not exist at lower levels. /28-29

- According to the systems view, the essential properties of an organism, or living system, are properties of the whole, which none of the parts have. They arise from the interactions and relationships among the parts. These properties are destroyed when the system is dissected, either physically or theoretically, into isolated elements. Although we can discern individual parts in any system, these parts are not isolated, and the nature of the whole is always different from the mere sum of its parts. /29

- The great shock of twentieth-century science has been that systems cannot be understood by analysis. The properties of the parts are not intrinsic properties but can be understood only within the context of the larger whole. Thus the relationship between the parts and the whole has been reversed. /29

- Analysis means taking something apart in order to understand it; systems thinking means putting it into the context of a larger whole. /30

- In the formalism of quantum theory these relationships are expressed in terms of probabilities, and the probabilities are determined by the dynamics of the whole system. Whereas in classical mechanics the properties and behavior of the parts determine those of the whole, the situation is reversed in quantum mechanics; it is the whole that determines the behavior of the parts. /31

- At the turn of the century, the philosopher Christian von Ehrenfels was the first to use Gestalt in the sense of an irreducible perceptual pattern, which sparked the school of Gestalt psychology. Ehrenfels characterized a gestalt by asserting that the whole is more than the sum of its parts, which would become the key formula of systems thinkers later on. /31

- The notion of pattern was always implicit in the writings of the Gestalt psychologists, who often used the analogy of a musical theme that can be played in different keys without losing its essential features. /32

- [T]he web of life consists of networks within networks. At each scale, under closer scrutiny, the nodes of the network reveal themselves as smaller networks. We tend to arrange these systems, all nesting within larger systems, in a hierarchical scheme by placing the larger systems above the smaller ones in pyramid fashion. But this is a human projection. In nature there is no 'above' or 'below,' and there are no hierarchies. There are only networks nesting within other networks. /35

- Ultimately—as quantum physics showed so dramatically—there are no parts at all. What we call a part if merely a pattern in an inseparable web of relationships. Therefore the shift from the parts to the whole can also be seen as a shift from objects to relationships. /37

- For thousands of years Western scientists and philosophers have used the metaphor of knowledge as a building, together with many other architectural metaphors derived from it. We speak of fundamental laws, fundamental principles, basic building blocks, and the like, and we assert that the edifice of science must be built on firm foundations. /38

- In the new systems thinking, the metaphor of knowledge as a building is being replaced by that of the network. As we perceive reality as a network of relationships, our descriptions, too, form an interconnected network of concepts and models in which there are no foundations. (...) When this approach is applied to science as a whole, it implies that physics can no longer be seen as the most fundamental level of science. Since there are no foundations in the network, the phenomena described by physics are not any more fundamental than those described by, say, biology or psychology. They belong to different systems levels, but none of those levels is any more fundamental than the others. /39

- The new paradigm implies that epistemology—understanding of the process of knowing—has to be included explicitly in the description of natural phenomena. /40

THE WEB OF LIFE

- When we draw a picture of a tree, most of us will not draw the roots. Yet the roots of a tree are often as expansive as the parts we see. In a forest, moreover, the roots of all trees are interconnected and form a dense underground network in which there are no precise boundaries between individual trees. In short, what we call a tree depends on our perceptions. It depends, as we say in science, on our methods of observation and measurement. In the words of Heisenberg: 'What we observe is not nature itself, but nature exposed to our method of questioning.' Thus systems thinking involves a shift from objective to 'epistemic' science, to a framework in which epistemology—'the method of questioning'—becomes an integral part of scientific theories. /40

- The criteria of systems thinking described in this brief summary are all interdependent. Nature is seen as an interconnected web of relationships, in which the identification of specific patterns as 'objects' depends on the human observer and the process of knowing. This web of relationships is described in terms of a / corresponding network of concepts and models, none of which is any more fundamental than the others. /40-41

- What makes it possible to turn the systems approach into a science is the discovery that there is approximate knowledge. This insight is crucial to all of modern science. The old paradigm is based on the Cartesian belief in the certainty of scientific knowledge. In the new paradigm it is recognized that all scientific concepts and theories are limited and approximate. Science can never provide any complete and definite understanding. /41

- Systems thinking is always process thinking. /42

- Unlike closed systems, which settle into a state of thermal equilibrium, open systems maintain themselves far from equilibrium in this 'steady state' characterized by continual flow and change. Bertalanffy coined the German term *Fliessgleichgewicht* ('flowing balance') to describe such a state of dynamic balance. He recognized clearly that classical thermodynamics, which deals with closed systems at or near equilibrium, is inappropriate to describe open systems in steady states far from equilibrium. /48

- Bertalanffy correctly identified the characteristics of the steady state as those of the process of metabolism, which led him to postulate self-regulation as another key property of open systems. This idea was refined by Prigogine thirty years later in terms of the self-organization of 'dissipative structures.' /49

- The cyberneticists were neither biologists nor ecologists; they were mathematicians, neuroscientists, social scientists, and engineers. They were concerned with a different level of description, concentrating on patterns of communication, especially in closed loops and networks. /51

- A feedback loop is a circular arrangement of causally connected elements, in which an initial cause propagates around the links of the loop, so that each element has an effect on the next, until the last 'feeds back' the effect into the first element of the cycle. /56

- Like the Cartesian model of the body as a clockwork, that of the brain as a computer was very useful at first, providing an exciting framework for a new scientific understanding of cognition and leading to many fresh avenues of research. By the mid-1960s, however, the original model, which encouraged the exploration of its own limitations and the discussion of alternatives, had hardened into a dogma, as so often happens in science. During the subsequent decade almost all of neurobiology was dominated by the information-processing perspective, whose origins and underlying assumptions were hardly ever questioned anymore. /67

- Recent developments in cognitive science have made it clear that human intelligence is utterly different from machine, or 'artificial,' intelligence. The human nervous system does not process any information (in the sense of discrete elements existing ready-made in the outside world, to be picked up by the cognitive system), but interacts with the environment by continually modulating its structure. Moreover, neuroscientists have discovered strong evidence that human intelligence, human memory, and human decisions are never completely rational but are always colored by emotions, as we all know from experience. /68

- Our thinking is always accompanied by bodily sensations and processes. Even if we often tend to suppress these, we always think also with our body; and since computers do not have such a body, truly human problems will always be foreign to their intelligence. /68

- Increasingly, all forms of culture are being subordinated to technology, and technological innovation, rather / than the increase in human well-being, has become synonymous with progress. The spiritual impoverishment and loss of cultural diversity through excessive use of computers is especially serious in the field of education. (…) The use of computers in school is based on the now outdated view of human beings as information processors, which continually reinforces erroneous mechanistic concepts of thinking, knowledge, and communication. Information is presented as the basis of thinking, whereas in reality the human mind thinks with ideas, not with information. /70

- In the computer model of cognition, knowledge is seen as context and value free, based on abstract data. But all meaningful knowledge is contextual knowledge, and much of it is tacit and experiential. Similarly, language is seen as a conduit through which 'objective' information is communicated. In reality, … language is metaphoric, conveying tacit understandings shared within a culture. /70

- Critical arguments had been presented already during the pioneering phase of cybernetics. For example, it was argued that in actual brains there are no rules; there is no central logical processor, and information is not stored locally. Brains seem to operate on the basis of massive connectivity, storing information distributively and manifesting a self-organizing capacity that is nowhere to be found in computers. However, these alternative ideas were eclipsed in favor of the dominant computational view, until they reemerged thirty years later during the 1970s, when systems thinkers became fascinated by a new phenomenon with an evocative name—self organization. /71

- T[he] triumph of molecular biology resulted in the widespread belief that all biological functions can be explained in terms of molecular structures and mechanisms. Thus most biologists have become fervent reductionists, concerned with molecular details. Molecular biology, originally a small branch of the life sciences, has now

- become a pervasive and exclusive way of thinking that has led to a severe distortion of biological research. /77

- It could be argued ... that the understanding of living organisms as energetically open but organizationally closed systems, the recognition of feedback as the essential mechanism of homeostasis, and the cybernetic models of neural processes—to name just three examples that were well established at the time—represented major advances in the scientific understanding of life. /78

- The cyberneticists concentrated on nonlinear phenomena like feedback loops and neural networks, and they had the beginnings of a corresponding nonlinear mathematics, but the real breakthrough came several decades later and was linked closely to the development of a new generation of powerful computers. /79

- To understand the phenomenon of self-organization, we first need to understand the importance of pattern. The idea of a pattern of organization—a configuration of relationships characteristic of a particular system—became the explicit focus of systems thinking in cybernetics and has been a crucial concept ever since. From the systems point of view, the understanding of life begins with the understanding of pattern. /80

- In the study of structure we measure and weigh things. Patterns, however, cannot be measured or weighed; they must be mapped. To understand a pattern we must map a configuration of relationships. In other words, structure involves quantities, while pattern involves qualities. /81

- Systemic properties are properties of pattern. What is destroyed when a living organism is dissected is its pattern. The components are still there, but the configuration of relationships among them—the pattern—is destroyed, and thus the organism dies. /81

- There is something else to life, something nonmaterial and irreducible—a pattern of organization. /81

- The structure of the human brain is enormously complex. It contains about 10 billion nerve cells (neurons), which are interlinked in a vast network through 1,000 billion junctions (synapses). The whole brain can be divided into subsections, or subnetworks, which communicate with each other in network fashion. All this results in intricate patterns of intertwined webs, networks nesting within larger networks. / 82

- Because networks of communication may generate feedback loops, they may acquire the ability to regulate themselves. For example, a community that maintains an active network of communication will learn from its mistakes, because the consequences of a mistake will spread through the network and return to the / source along feedback loops. Thus the community can correct its mistakes, regulate itself, and organize itself. Indeed, self-organization has emerged as perhaps the central concept in the systems view of life, and like the concepts of feedback and self-regulation, it is linked closely to networks. The pattern of life, we might say, is a network pattern capable of self-organization. This is a simple definition, yet it is based on recent discoveries at the very forefront of science. / 83

- Hypercycles turn out to be not only remarkably stable, but also capable of self-replication and of correcting replication error, which means that they can conserve and transmit complex information. (…) The lesson to be learned here seems to be that the roots of life reach down into the realm of nonliving matter. / 94

- Lovelock recognized the Earth's atmosphere as an open system, far from equilibrium, characterized by a constant flow of energy and matter. / 102

- The new mathematics … is one of relationships and patterns. It is qualitative rather than quantitative and thus embodies the shift of emphasis that is characteristic of systems thinking—from objects to relationships, from quantity to quality, from substance to pattern. / 113

- Nonlinear phenomena dominate much more of the inanimate world than we had thought, and they are an essential aspect of the network patterns of living systems. Dynamical systems theory is the first mathematics that enables scientists to deal with the full complexity of these nonlinear phenomena. / 123

- The exploration of nonlinear systems over the past decades has had a profound impact on science as a whole, as it has forced us to reevaluate some very basic notions about the relationships between a mathematical model and the phenomena it describes. One of those notions concerns our understanding of simplicity and complexity. /123

- The behavior of chaotic systems is not merely random but shows a deeper level of patterned order. /123

- In nonlinear systems ... small changes may have dramatic effects because they may be amplified repeatedly by self-reinforcing feedback. Such nonlinear feedback processes are the basis of the instabilities and the sudden emergence of new forms of order that are so characteristic of self-organization. /124

- The Mandelbrot set is a storehouse of patterns of infinite detail and variations. Strictly speaking, it is not self-similar because it not only repeats the same patterns over and over again, including small replicas of the entire set, but also contains elements from an infinite number of Julia sets! It is thus a 'superfractal' of inconceivable complexity. /151

- T[he] isolation of mathematics is a striking sign of our intellectual fragmentation and as such is a relatively recent phenomenon. Throughout the centuries many of the great mathematicians made outstanding contributions to other fields as well. In the eleventh / century the Persian poet Omar Khyyám, who is world renowned as the author of the Rubáiyát, also wrote a pioneering book on algebra and served as the official astronomer at the caliph's court. Descartes, the founder of modern philosophy, was a brilliant mathematician and also practiced medicine. Both inventors of the differential calculus, Newton and Leibniz, were active in many fields besides mathematics. Newton was a 'natural philosopher' who made fundamental contributions to virtually all branches of science that were known at his time, in addition to studying alchemy, theology, and history. Leibniz is known primarily as a philosopher, but he was also the founder of symbolic logic and was active as a diplomat and historian during most of his life. The great mathematician Gauss was also a physicist and astronomer, and he invented several useful instruments, including the electric telegraph. /151-152

- The great fascination exerted by chaos theory and fractal geometry on people in all disciplines—from scientists to managers to artists—may indeed be a hopeful sign that the isolation of mathematics is ending. Today the new mathematics of complexity is making more and more people realize that mathematics is much more than dry formulas; that the understanding of pattern is crucial to understand the living world around us; and that all questions of pattern, order, and complexity are essentially mathematical. /152-153

- From Pythagoras to Aristotle, to Goethe, and to the organismic biologists, there is a continuous intellectual tradition that struggles with the understanding of pattern, realizing that it is crucial to the understanding of living form. /158

- The understanding of pattern, then, will be of crucial importance to the scientific understanding of life ./158

- T[he] striking property of living systems suggests process as a third for a comprehensive description of the nature of life. The process of life is the activity involved in the continual embodiment of the system's pattern of organization. Thus the process criterion is the link between pattern and structure. /168-169

- The pattern of organization can be recognized only if it is embodied in a physical structure, and in living systems this embodiment is an ongoing process. Thus structure and process are inextricably linked. One could say that the three criteria—pattern, structure, and process—are three different but inseparable perspectives of the phenomenon of life. /160

- To find out whether a particular system—a crystal, a virus, a cell, or the planet Earth—is alive, all we need to do is to find out whether the pattern of organization is that of an autopoietic network. If it is, we are dealing with a living system; if it is not, the system is nonliving. /161

- Autopoiesis and cognition are two different aspects of the same phenomenon of life. In the new theory all living systems are cognitive systems, and cognition always implies the existence of an autopoietic network. /161

- Autopoiesis, or 'self-making,' is a network pattern in which the function of each component is to participate in the production or transformation of other components of the network. In this way the network continually makes itself. It is produced by its components and in turn produces those components. / 162

- Since all components of an autopoietic network are produced by other components in the network, the entire system is organizationally closed, even though it is open with regard to the flow of energy and matter. This organizational closure implies that a living system is self-organizing in the sense that its order and behavior are not imposed by the environment but are established by the system itself. In other words, living systems are autonomous. This does not mean that they are isolated from their environment. On the contrary, they interact with the environment through a continual exchange of energy and matter. / 167

- Autopoiesis, then, is seen as the pattern underlying the phenomenon of self-organization, or autonomy, that is so characteristic of all living systems. / 168

- [A]utopoietic networks must continually regenerate themselves to maintain their organization. This, of course, is a well-known characteristic of life. / 168

- According to the theory of living systems, mind is not a thing but a process—the very process of life. In other words, the organizing activity of living systems, at all levels of life, is mental activity. The interactions of a living organism—plant, animal, or human—with its environment are cognitive, or mental interactions. Thus life and cognition become inseparably connected. Mind—or, more accurately, mental process—is immanent in matter at all levels of life. /172

- The central insight of the Santiago theory is the same as Bateson's—the identification of cognition, the process of knowing, with the process of life. / 174

- The new concept of cognition, the process of knowing, is thus much broader than that of thinking. It involves perception, emotion, and action—the entire process of life. In the human realm cognition also includes language, conceptual thinking, and all the other attributes of human consciousness. The general concept, however, is much broader and does not necessarily involve thinking. /175

- The Santiago theory provides ... the first coherent scientific framework that really overcomes the Cartesian split. Mind and matter no longer appear to belong to two separate categories but are seen as representing merely different aspects, or dimensions, of the same phenomenon of life. /175

- In the Santiago theory the relationship between mind and brain is simple and clear. Descartes' characterization of mind as 'the thinking thing' (res cogitans) is finally abandoned. Mind is not a thing but a process—the process of cognition, which is identified with the process of life. The brain is a specific structure through which this process operates. The relationship between mind and brain, therefore, is one between process and structure. /175

- The brain is, of course, not the only structure through which the process of cognition operates. The entire dissipative structure of the organism participates in the process of cognition, whether or not the organism has a brain and a higher nervous system. /175-176

- [A]ll organisms in an ecosystem produce wastes, but what is waste for one species is food for another, so that wastes are continually recycled and the ecosystem as a whole generally remains without waste. /177

- By blending water and minerals from below with sunlight and CO_2 from above, green plants link the earth and the sky. We tend to believe that plants grow out of the soil, but in fact most of their substance comes from the air. The bulk of the cellulose and the other organic compounds produced through photosynthesis consists of heavy carbon and oxygen atoms, which plants take directly from the air in the form of CO_2. Thus the weight of a wooden log comes almost entirely from the air. When we burn a log in a fireplace, oxygen and carbon combine once more into CO_2, and in the light and heat of the fire we recover part of the solar energy that went into making the wood. /178

- Prigogine realized that classical thermodynamics, the first science of complexity, is inappropriate to describe systems far from equilibrium because of the linear nature of its mathematical structure. /181

- Farther away from equilibrium, the fluxes are stronger, entropy production increases, and the system no longer tends toward equilibrium. On the contrary, it may encounter instabilities leading to new forms of order that move the system farther and farther away from the equilibrium state. In other words, far from equilibrium, dissipative structures may develop into forms of ever-increasing complexity. /181

- Far from equilibrium, the system's flow processes are interlinked through multiple feedback loops, and the corresponding mathematical equations are nonlinear. The farther a dissipative structure is from equilibrium, the greater is its complexity and the higher is the degree of nonlinearity in the mathematical equations describing it. (…) Nonlinear equations usually have more than one solution; the higher the nonlinearity, the greater the number of solutions. This means that new situations may emerge at any moment. Mathematically speaking, the system encounters a bifurcation point in such a case, at which it may branch off into an entirely new state. /182

- Near equilibrium we find repetitive phenomena and universal laws. As we move away from equilibrium, we move from the universal to the unique, toward richness and variety. /182

- In a Newtonian world there would be no chemistry and no life. /184

- T[he] new perception of order and disorder represents an inversion of traditional scientific views. According to the classical view, for which physics was the principal source of concepts and metaphors, order is associated with equilibrium, as, for example, in crystals and other static structures, and disorder with non-equilibrium situations, such as turbulence. In the new science of complexity, which takes its inspiration from the web of life, we learn that nonequilibrium is a source of order. The turbulent flows of water and air, while appearing chaotic, are really highly organized, exhibiting complex patterns of vortices dividing and subdividing again at smaller and smaller scales. In living systems the order arising from nonequilibrium is far more evident, being manifest in the richness, diversity, and beauty of life all around us. Throughout the living world chaos is transformed into order. /190

- The conceptual shift implied in Prigogine's theory involves several closely related ideas. The description of dissipative structures that exist far from equilibrium requires a nonlinear mathematical formalism, capable of modeling multiple interlinked feedback loops. In living organisms these are catalytic loops (that is, nonlinear, irreversible chemical processes), which lead to instabilities through repeated self-amplifying feedback. When a dissipative structure reaches such a point of instability, called a bifurcation point, an element of indeterminacy enters into the theory. At the bifurcation point the system's behavior is inherently unpredictable. In particular, new structures of higher order and complexity may emerge spontaneously. Thus self-organization, the spontaneous emergence of order, results from the combined effects of non-equilibrium, irreversibility, feedback loops, and instability. /192

- Instead of being a machine, nature at large turns out to be more like human nature—unpredictable, sensitive to the surrounding world, influenced by small fluctuations. Accordingly the appropriate way of approaching nature to learn about her complexity and beauty is not through domination and control, but through respect, cooperation, and dialogue. /193

- In the deterministic world of Newton there is no history and no creativity. In the living world of dissipative structures history plays an important role, the future is uncertain, and thus uncertainty is at the heart of creativity. /193

- Many of these cyclical changes occur much faster than one / would imagine. For example. Our pancreas replaces most of its cells every twenty-four hours, the cells of our stomach lining are reproduced every three days, our white blood cells are renewed in ten days, and 98 percent of the protein in our brain is turned over in less than one month. Even more amazing, our skin replaces its cells at the rate of one hundred thousand cells per minute. In fact, most of the dust in our homes consists of dead skin cells. /219

- Kicking a stone and kicking a dog are two very different stories, as Gregory Bateson was fond of pointing out. The stone will react to the kick according to a linear chain of cause and effect. Its behavior can be calculated by applying the basic laws of Newtonian mechanics. The dog will respond with structural changes according to its own nature and (nonlinear) pattern of organization. The resulting behavior is generally unpredictable. /219

- Like Prigogine's theory of dissipative structures, the theory of Autopoiesis shows that creativity—the generation of configurations that are constantly new—is a key property of all living systems. A special form of this creativity is the generation of diversity through reproduction, from simple cell division to the highly complex dance of sexual reproduction. For most living organisms ontogeny is not a linear path of development but a cycle, and reproduction is a vital step in that cycle. /221

- Rather than seeing evolution as the result of random mutations and natural selection, we are beginning to recognize the creative unfolding of life in forms of ever-increasing diversity and complexity as an inherent characteristic of all living systems. Although mutation and natural selection are still acknowledged as important aspects of biological evolution, the central focus is on creativity, on life's constant reaching out into novelty. /222

- Fast bacteria can divide about every twenty minutes, so that in principle several billion individual bacteria can be generated from a single cell in less than a day. /228

- Bacteria are able to adapt to environmental changes in a few years, where larger organisms would need thousands of years of evolutionary adaptation. /229

- In other words, all bacteria are part of a single microscopic web of life. /230

- The most striking evidence for evolution through symbiosis is presented by the so-called mitochondria, the 'powerhouses' inside most nucleated cells. These vital parts of all animal and plant cells, which carry out cellular respiration, contain their own genetic material and

> reproduce independently and at different times from the rest of the cell. /231

- The evolutionary unfolding of life over billions of years is a breathtaking story. Driven by the creativity inherent in all living systems, expressed through three distinct avenues—mutations, the trading of genes, and symbioses—and honed by natural selection, the planet's living patina expanded and intensified in forms of ever-increasing diversity. /232

- [A]bout 3.5 billion years ago, the first autopoietic bacterial cells were born, and the evolution of life began. /236

- Perhaps the most important task was to develop a variety of new metabolic pathways for extracting food and energy from the environment. One of the first bacterial inventions was fermentation—the breaking down of sugars and conversion into ATP molecules, the 'energy carriers' that fuel all cellular processes. This innovation allowed the fermenting bacteria to live off chemicals in the earth, in mud and water, protected from the harsh sunlight. /236

- During subsequent stages of evolution, the microorganisms formed alliances and coevolved with plants and animals, and today our environment is so interwoven with bacteria that it is almost impossible to say where the inanimate world ends and life begins. We tend to associate bacteria with disease, but they are also vital for our survival, as they are for the survival of all animals and plants. /239

- The recognition of symbiosis as a major evolutionary force has profound philosophical implications. All larger organisms, including ourselves, are living testimonies to the fact that destructive practices do not work in the long run. In the end the aggressors always destroy themselves, making way for others who know how to cooperate and get along. Life is much less a competitive struggle for survival than the triumph of cooperation and creativity. Indeed, since the creation of the first nucleated cells, evolution has proceeded through ever more intricate arrangements of cooperation and coevolution. /243

- Like so many other life processes, rapid motion was invented by bacteria. The fastest member of the microcosm is a tiny, hairlike creature called spirochete ('coiled hair'), also known as the 'corkscrew bacterium,' which spirals in rapid motion. By attaching themselves symbiotically to larger cells, the rapidly moving corkscrew bacteria gave those cells the tremendous advantages of locomotion—the ability to avoid danger and seek out food. Over time the corkscrew bacteria progressively lost their distinct traits and evolved into the well-known 'cell whips'—flagellae, cilia, and the like—that propel a wide variety of nucleated cells with undulating or whipping motions. /244

- As a scientific hypothesis the concept of symbiogenesis—the creation of new forms of life through the merging of different species—is barely thirty years old. But as a cultural myth the idea seems to be as old as humanity itself. Religious epics, legends, fairy tales, and other mythical stories around the world are full of fantastic creatures—sphinxes, mermaids, griffons, centaurs, and more—born from the blending of two or more species. Like the new eukaryotic cells, these creatures are made of components that are entirely familiar, but their combinations are novel and startling. /244-245

- Depictions of these hybrid beings are often frightening, but many of them, curiously, are seen as bearers of good fortune. For example, the god Ganesha, who has a human body with an elephant head, is one of the most revered deities in India, worshiped as a symbol of good luck and a helper in overcoming obstacles. Somehow the collective human unconscious seems to have known from ancient times that long-term symbioses are profoundly beneficial for all life. /245

- As the specialization of cells continued in larger and more complex forms of life, the capability of self-repair and regeneration diminished progressively. Flatworms, polyps and starfish can regenerate almost their entire bodies from small fractions; lizards, salamanders, crabs, lobsters, and many insects are still able to grow back lost organs or limbs; but in higher animals regeneration is limited to renewing tissues in the healing of injuries. /246

- Among the many multicellular organizations that evolved out of tightly knit communities of microorganisms, three—plants, fungi, and animals—have been so successful in reproducing, diversifying, and expanding, and expanding over the Earth that they are classified by biologists as 'kingdoms', the broadest category of living organisms. All in all there are five of these kingdoms—bacteria (microorganisms without cell nuclei), protists (microorganisms with nucleated cells), plants, fungi, and animals. Each of the kingdoms is divided into a hierarchy of subcategories, or taxa, beginning with phylum and ending with genus and species. /247

- In the emerging theory of living systems mind is not a thing, but a process. It is cognition, the process of knowing, and it is identified with the process of life itself./264

- The identification of mind, or cognition, with the process of life is a radically new idea in science, but it is also one of the deepest and most archaic intuitions of humanity. In ancient times the rational human mind was seen as merely one aspect of the immaterial soul, or spirit. /264

- [T]he etymological roots of 'soul' and 'spirit' mean breath in many antique languages. The words for 'soul' in Sanskrit (atman), Greek (pneuma), and Latin (anima) all mean 'breath.' The same is true of the world for 'spirit' in Latin (spiritus), in Greek (psyche), and in Hebrew (ruah). /264

- The computer model of cognition was finally subjected to serious questioning in the 1970's when the concept of self-organization emerged. (...) These observations suggested a shift of focus—from symbols to connectivity, from local rules to global coherence, from information processing to the emergent properties of neural networks. /266

- The range of interactions a living system can have with its environment defines its 'cognitive domain.' Emotions are an integral part of this domain. For example, when we respond to an insult by getting angry, that entire pattern of physiological processes—a red face, faster breathing, trembling, and so on—is part of cognition. In fact, recent research strongly indicates that there is an emotional coloring to every cognitive act. /269

- According to the Santiago theory, cognition is not a representation of an independent, pregiven world, but rather a bringing forth of a world. What is brought forth by a particular organism in the process of living is not the world but a world, one that is always dependent upon the organism's structure. Since individual organisms within a species have more or less the same structure, they bring forth similar worlds. We humans, moreover, share an abstract world of language and thought through which we bring forth our world together. /270

- A computer processes information, which means that it manipulates symbols based on certain rules. The symbols are distinct elements fed into the computer from outside, and during the information processing there is no change in the structure of the machine. The physical structure of the computer is fixed, determined by its design and construction. The nervous system of a living organism ... interacts with its environment by continually modulating its structure, so that at any moment its physical / structure is a record of previous structural changes. The nervous system does not process information from the outside world but, on the contrary, brings forth a world in the process of cognition. /274-275

- A lot of confusion is caused by the fact that computer scientists use words such as 'intelligence,' 'memory,' and 'language' to describe computers, thus implying that these expressions refer to the human phenomena we know well from experience. This is a serious misunderstanding. For example, the very essence of intelligence is to act appropriately when a problem is not clearly defined and solutions are not evident. Intelligent human behavior in such situations is based on common sense, accumulated from lived experience. Common sense, however, is not available to computers because of their blindness of abstraction and the intrinsic limitations / of formal operations, and therefore it is impossible to program computers to be intelligent. /275-276

- The reason is that language is embedded in a web of social and cultural conventions that provides an unspoken context of meaning. We understand this context because it is common sense to us, but a computer cannot be programmed with common sense and therefore does not understand language. /276

- Mind is not a thing but a process—the process of cognition, which is identified with the process of life. The brain is a specific structure through which this process operates. Thus the relationship between mind and brain is one between process and structure. /278

- Recent research has shown that under normal conditions the antibodies circulating in the body bind to many (if not all) types of cells, including themselves. The entire system looks much more like a network, more like people talking to each other, than soldiers out looking for an enemy. Gradually immunologists have been forced to shift their perception from an immune system to an immune network. /279

- Rather than merely reacting against foreign agents, the immune system serves the important function of regulating the organism's cellular and molecular repertoire. As Francisco Varela and immunologist Antonio Coutinho explain, 'The mutual dance between immune system and body ... allows the body to have a changing and plastic identity throughout its life and its multiple encounters.' /280

- When immunologists inject large amounts of a foreign agent into the body, as they do in standard animal experiments, the immune system reacts with the massive defense response described in the classical theory. However, as Varela and Coutinho point out, this is a highly contrived laboratory situation. In its natural surroundings an animal does not receive large amounts of harmful substances. The small amounts that do enter its body are incorporated naturally into the ongoing regulatory activities of its immune network. /280

- Defensive immune activity is very important, but in the new view it is a secondary effect of the much more central cognitive activity of the immune system, which maintains the body's molecular identity. /281

- The nervous system, consisting of the brain and of a network of nerve cells throughout the body, is the seat of memory, thought, and emotion. The endocrine system, consisting of the glands and the hormones, is the body's main regulatory system, controlling and integrating various bodily functions. The immune system, consisting of the spleen, the bone marrow, the lymph nodes, and the immune cells circulating through the body, is the body's defense system, responsible for tissue integrity and controlling wound healing and tissue-repair mechanisms. In accord with this separation the three systems are studied in three separate disciplines—neuroscience, endocrinology, and immunology. However, the recent peptide research has shown in dramatic ways that these conceptual separations are merely historical artifacts that can no longer be maintained. According to Candace Pert, the three systems must be seen as forming a single psychosomatic network. /282

- Peptides are the biochemical manifestation of emotions; they play a crucial role in the coordinating activities of the immune system; they interlink and integrate mental, emotional, and biological activities. /283

- The entire group of sixty to seventy peptides may constitute a universal biochemical language of emotions. /284

- Traditionally neuroscientists have associated emotions with specific areas in the brain, notably the limbic system. This is indeed correct. The limbic system turns out to be highly enriched with peptides. However, it is not the only part of the body where peptide receptors are concentrated. For example, the entire intestine is lined with peptide receptors. This is why we have 'gut feelings.' We literally feel our emotions in our gut. /284

- The uniqueness of being human lies in our ability to continually weave the linguistic network in which we are embedded. To be human is to exist in language. In language we coordinate our behavior, and together in language we bring forth our world. /290

- From the perspective of the Santiago theory, the currently fashionable attempts to explain human consciousness in terms of quantum effects in the brain or other neurophysiological processes are all bound to fail. Self-awareness and the unfolding of our inner world of concepts and ideas are not only inaccessible to explanations in terms of physics and chemistry; they cannot even be understood through the biology or psychology of a single organism. According to Maturana, we can understand human consciousness only through language and the whole social context in which it is embedded. As its Latin root—*conscire* ('knowing together')—might indicate, consciousness is essentially a social phenomenon. /291

- Over the centuries the word maya—one of the most important terms in Indian philosophy—changed its meaning. From the creative power of Brahman it came to signify the psychological state of anybody under the spell of the magic play. As long as we confuse the material forms of the play with objective reality, without perceiving the unity of Brahman underlying all these forms, we are under the spell of maya. /291

- The fact that neural circuits tend to oscillate rhythmically is well-known to neuroscientists, and recent research has shown that these oscillations are not restricted to the cerebral cortex but occur at various levels in the nervous system. /293

- According to Varela, the primary conscious experience, common to all higher vertebrates, is not located in a specific part of the brain, nor can it be identified in terms of specific neural structures. It is the manifestation of a particular cognitive process—a transient synchronization of diverse, rhythmically oscillating neural circuits. /293

- The power of abstract thinking has led us to treat the natural environment—the web of life—as if it consisted of separate parts, to be exploited by different interest groups. Moreover, we have extended this fragmented view to our human society, dividing it into different nations, races, religious and political groups. The belief that all these fragments—in ourselves, in our environment, and in our society—are really separate has alienated us from nature and from our fellow human beings and thus has diminished us. To regain our full humanity, we have to regain our experience of connectedness with the entire web of life. This reconnecting, *religio* in Latin, is the very essence of the spiritual grounding of deep ecology. /296

- Ecosystems differ from individual organisms in that they are largely (but not completely) closed systems with respect to the flow of matter, while being open with respect to the flow of energy. /299

- The 1991 war in the Persian Gulf, for example, which killed hundreds of thousands, impoverished millions, and caused unprecedented environmental disasters, had its roots to a large extent in the misguided energy policies of the Reagan and Bush administrations. /299-300

- To describe solar energy as economically efficient assumes that the costs of energy production are counted honestly. This is not the case in most of today's market economies. The so-called free market does not provide consumers with proper information, because the social and environmental costs of production are not part of current economic models. These costs are labeled 'external' variables by corporate and government economics, because they do not fit into their theoretical framework. /300

- Corporate economists treat as free commodities not only the air, water, and soil, but also the delicate web of social relations, which is severely affected by continuing economic expansion. Private profits are being made at public costs in the deterioration of the environment and general quality of life, and at the expense of future generations. The marketplace simply gives us the wrong information. There is a lack of feedback, and basic ecological literacy tells us that such a system is not sustainable. /300

- One of the most effective ways to change the situation would be an ecological tax reform. Such a tax reform would be strictly revenue neutral, shifting the tax burden from income taxes to 'eco-taxes.' /300

- Partnership is an essential characteristic of sustainable communities. The cyclical exchanges of energy and resources in an ecosystem are sustained by pervasive cooperation. Indeed, we have seen that since the creation of the first nucleated cells over two billion years ago, life on Earth has proceeded through ever more intricate arrangements of cooperation and coevolution. Partnership—the tendency to associate, establish links, live inside one another, and cooperate—is one of the hallmarks of life. /301

- The flexibility of an ecosystem is a consequence of its multiple feedback loops, which tend to bring the system back into balance whenever there is a deviation from the norm, due to changing environmental conditions. For example, if an unusually warm summer results in increased growth of algae in a lake, some species of fish feeding on these algae may flourish and breed more, so that their numbers increase and they begin to deplete the algae. Once their major source of food is reduced, the fish will begin to die out. As the fish population drops, the algae will recover and expand again. In this way the original disturbance generates a fluctuation around a feedback loop, which eventually brings the fish/algae system back into balance. /302

- All ecological fluctuations take place between tolerance limits. There is always the danger that the whole system will collapse when the fluctuation goes beyond those limits and the system can no longer compensate for it. The same is true for human communities. Lack of flexibility manifests itself as stress. In particular, stress will occur when one or more variables of the system are pushed to their extreme values, which induces increased rigidity throughout the system. Temporary stress is an essential aspect of life, but prolonged stress is harmful and destructive to the system. These considerations lead to the important realization that managing a social system—a company, a city, or an economy—means finding the optimal values for the system's variables. If one tries to maximize any single variable instead of optimizing it, this will invariably lead to the destruction of the system as a whole. /302-303

- A diverse ecosystem will also be resilient, because it contains many species with overlapping ecological functions that can partially replace one another. /303

- Diversity means many different relationships, many different approaches to the same problem. A diverse community is a resilient community, capable of adapting to changing situations. /303

Chapter Nine
The Hidden Connections

The Hidden Connections

A Science for Sustainable Living
New York: Anchor Books, 2004
Author Copyright 2002

Review

The Hidden Connections is perhaps the most lucid of Capra's books. This being said, I could well imagine that if you begin reading Capra with the present book, without reading his previous books first, you might get stuck somewhere in the midst of it—simply because you lack out on essential information that is contained in Capra's earlier books.

At the very onset of *The Hidden Connections*, Capra reveals an important detail about himself and his unusual development as a scientist:

> My extension of the systems approach to the social domain explicitly includes the material world. This is unusual, because traditionally social scientists have not been very interested in the world of matter. Our academic disciplines have been organized in such a way that the natural sciences deal with material structures while the social sciences deal with social structures, which are understood to be, essentially, rules of behavior. In the future, this strict division will no longer be possible, because the key challenge of this new century—for social scientists, natural scientists and everybody else—will be to build ecologically sustainable communities, designed in such a way that their technologies and social institutions—their material and social structures—do not interfere with nature's inherent ability to sustain life./xix

Capra starts, systemically sound, with the cell, noting that the simplest living system is the cell, and especially, the bacterial cell. Then Capra looks at what *membranes* are, and what they do, and this is highly revealing, and teaches an important lesson about relationships. I haven't found this insightful metaphor anywhere else, and it showed me right at the start of this book that it's going to be highly substantial lecture:

> A membrane is very different from a cell wall. Whereas cell walls are rigid structures, membranes are always active, opening and closing continually, keeping certain substances out and letting others in./8

> The cell's metabolic reactions involve a variety of ions, and the membrane, by being semipermeable, controls their proportions and keeps them in balance. Another critical activity of the membrane is to continually pump out excessive calcium waste, so that the calcium remaining within the cell is kept at the precise, very low level required for its metabolic functions. All these activities help to maintain the cell as a distinct entity and protect it from harmful environmental influences. Indeed, the first thing a bacterium does when it is attacked by another organism is to make membranes. /Id.

The next important point to understand how nature 'thinks' is the cell's metabolism, the network that serves recycling. Capra succinctly elaborates:

> When we take a closer look at the processes of metabolism, we notice that they form a chemical network. This is another fundamental feature of life. As ecosystems are understood in terms of food webs (networks of organisms), so organisms are viewed as networks of cells, organs and organ systems, and cells as networks of molecules. One of the key insights

> of the systems approach has been the realization that the network is a pattern that is common to all life. Wherever we see life, we see networks. (...) The metabolic network of a cell involves very special dynamics that differ strikingly from the cell's nonliving environment. Taking in nutrients from the outside world, the cell sustains itself by means of a network of chemical reactions that take place inside the boundary and produce all of the cell's components, including those of the boundary itself./9

I shall leave out in this review the long passages in which Capra explains the essential contributions of systems researchers such as Humberto Maturana, Francisco Varela, or Ilya Prigogine, as this would render this review definitely too extensive. I shall thus restrict myself to a few remarks for describing the core of systems research that Capra unfolds in this book:

> The starting point for this is the observation that all cellular structures exist far from equilibrium state—in other words, the cell would die—if the cellular metabolism did not use a continual flow of energy to restore structures as fast as they are decaying. This means that we need to describe the cell as an open system. Living systems are organizationally closed—they are autopoietic networks—but materially and energetically open./13

One of the most important insights we gain from systems theory and the close observation of natural processes is the relationship between chaos and order. What is chaos? What is order? We all have some preconceptions here. Sure, but I promise you that when you read this book, you will let go all of them, because they are wrong!

Chaos is not random, but ordered chaos, but order is not a stable condition. You may remember that we briefly discussed earlier on what self-organization means relating to systems. Here, Capra explains in more detail what self-organization actually does:

> Th[e] spontaneous emergence of order at critical points of instability is one of the most important concepts of the new understanding of life. It is technically known as self organization and is often referred to simply as emergence. It has been recognized as the dynamic origin of development, learning and evolution. In other words, creativity—the generation of new forms—is a key property of all living systems. And since emergence is an integral part of the dynamics of open systems, we reach the important conclusion that open systems develop and evolve. Life constantly reaches out into novelty./14

The next great error most of us are caught in is the discrimination between humans and animals when it is about cognition. Fact is that humans are not much more intelligent than Gorillas, only a little more, to be precise: we are 1.6 times more intelligent than gorillas. Besides that, it was believed that in animals cognition was working in basically different ways than in humans. This seems to have been an error. Researchers found you can talk with chimpanzees if you learn their language, and they can learn ours. Capra summarizes this research shortly:

> The unified, post-Cartesian view of mind, matter, and life also implies a radical reassessment of the relationships between humans and animals. Throughout most of Western philosophy, the capacity to reason was seen as a uniquely human characteristic, distinguishing us from all other animals. The communication studies with chimpanzees / have exposed the fallacy of this belief in the most dramatic of ways. They make it clear that the cognitive and emotional lives of animals and humans differ only by degree; that life is a great continuum in which differences between species are gradual and evolutionary. / 65-66

I shall finalize this review with some very interesting political and social hidden connections that Capra unveils in his book.

There are probably still people around who are fond of biotechnology, but I guess they just ignore the facts, and their knowledge is for the most part taken from the huge amount of propaganda material. Was it only for this enlightening information, the present book is worth its price as it daringly unveils the hidden facts and tells the truth!

> The most widespread use of plant biotechnology has been to develop herbicide-tolerant crops in order to boast the sales of particular herbicides. There is a strong likelihood that the transgenic plants will cross-pollinate with wild relatives in their surroundings, thus creating herbicide-resistant superweeds. Evidence indicates that such gene flows between transgenic crops and wild relatives are already occurring. / 193

Why do we need biotechnology? I guess certain people, corporations and their consorts need it for making huge amounts of money. But is it tolerable in a democracy that all suffer from the side effects of technologies that enrich a few? I learnt as a law student that such a kind of system is called an *oligarchy*, the reign of an elite. So I am seriously asking how we ever came to say that we are living in a democracy?

> In the animal kingdom, where cellular complexity is much higher, the side effects in genetically modified species are much worse. 'Super-salmon' which were engineered to grow as fast as possible, ended up with monstrous heads and died from not being able to breathe or feed properly. Similarly, a superpig with a human gene for a growth hormone turned out ulcerous, blind, and impotent. (...) The most horrifying and by now best-known story is probably that of the genetically altered hormone called recombinant bovine growth hormone, which has been used to stimulate milk production in cows despite the fact that American dairy farmers have produced vastly more milk than people can consume for the past fifty years. The effects of this genetic engineering folly on the cow's health are serious. They include bloat, diarrhea, diseases of the knees and feet, cystic ovaries, and many more. Besides, their milk may contain a substance that has been implicated in human breast and stomach cancers./198

Why do we need superpigs? It seems to me that they are the result of *quantitative* thinking, a primacy of quantity over quality, and this for the obvious reason of maximizing profits. This is a good example for the fact that we live in what has been called *the corporate society*, as the prototype of a society in which major corporations dictate the standards the government is going to follow and to enact as laws. Capra notes the details:

> In the United States, the biotech industry has persuaded the Food and Drug Administration (FDA) to treat GM food as substantially equivalent to traditional food, which allows food producers to evade normal testing by the FDA and the Environmental Protection Agency (EPA), and also leaves it to the companies' own discretion as to whether to label their products as genetically modified. Thus, the public is kept unaware of the rapid spread of transgenic foods and scientists will find it much harder to trace harmful effects. Indeed, buying organic is now the only way to avoid GM foods./199

In Germany, France, and most other European countries, the laws are different regarding genetically modified food. Capra informs:

> The governments of France, Italy, Greece, and Denmark announced that they would block the approval of new GM crops in the European Union. The European Commission made the labeling of GM foods mandatory, as did the governments of Japan, South Korea, Australia, and Mexico. In January 2000, 130 nations signed the groundbreaking Cartagena Protocol on Biosafety in Montreal, which gives nations the right to refuse entry to any genetically modified forms of life, despite vehement opposition from the United States. /228

As a trained lawyer, I can clearly see that we are facing currently a challenge to legally codify these new technologies—lest, as it were, they are going to codify us, entraining us in a turbulence of *faits établis*, and then the law will leap behind the actual developments. But the law should better accompany the research step by step so as to be updated with the explosive growth of these very heavily funded research disciplines. Capra writes:

> The development of such new biotechnologies will be a tremendous intellectual challenge, because we still do not understand how nature developed technologies during billions of years of evolution that are far superior to our human designs. How do mussels produce glue that sticks to anything in water? How do spiders spin a silk thread that, ounce for ounce, is five times stronger than steel? How do abalone grow a shell that is twice as tough as our high-tech ceramics? How do these creatures manufacture their miracle materials in water, at room temperature, silently, and without any toxic byproducts? / 204

Quotes

- Since we know that all living organisms are either single cells or multicellular, we know that the simplest living system is the cell. More precisely, it is a bacterial cell. / 4

- Since its beginning, life on earth has been associated with water. Bacteria move in water, and the metabolism inside their membranes takes place in a watery environment. In such fluid surroundings, a cell could never persist as a distinct entity without a physical barrier against free diffusion. The existence of membranes is therefore an essential condition for cellular life. / 8

- A membrane is very different from a cell wall. Whereas cell walls are rigid structures, membranes are always active, opening and closing continually, keeping certain substances out and letting others in. The cell's metabolic reactions involve a variety of ions, and the membrane, by being semipermeable, controls their proportions and keeps them in balance. Another critical activity of the membrane is to continually pump out excessive calcium waste, so that the calcium remaining within the cell is kept at the precise, very low level required for its metabolic functions. All these activities help to maintain the cell as a distinct entity and protect it from harmful environmental influences. Indeed, the first thing a bacterium does when it is attacked by another organism is to make membranes. /8

- The cell does not contain several distinct membranes, but rather has one single, interconnected membrane system. This so-called 'endomembrane system' is always in motion, wrapping itself around all the organelles and going out to the edge of the cell. It is a moving 'conveyor belt' that is continually produced, broken down and produced again. /8

- When we take a closer look at the processes of metabolism, we notice that they form a chemical network. This is another fundamental feature of life. As ecosystems are understood in terms of food webs (networks of organisms), so organisms are viewed as networks of cells, organs and organ systems, and cells as networks of molecules. One of the key insights of the systems approach has been the realization that the network is a pattern that is common to all life. Wherever we see life, we see networks. /9

- The metabolic network of a cell involves very special dynamics that differ strikingly from the cell's nonliving environment. Taking in nutrients from the outside world, the cell sustains itself by means of a network of chemical reactions that take place inside the boundary and produce all of the cell's components, including those of the boundary itself. /9

- The function of each component in this network is to transform or replace other components, so that the entire network continually generates itself. This is the key to the systemic definition of life: living networks continually create, or re-create, themselves by transforming or replacing their components. In this way they undergo continual structural changes while preserving their weblike patterns of organization. /9-10

- The dynamic of self-generation was identified as a key characteristic of life by biologists Humberto Maturana and Francisco Varela, who gave it the name 'autopoiesis' (literally, 'self-making'). The concept of autopoiesis combines the two defining characteristics of cellular life mentioned above, the physical boundary and the metabolic network. Unlike the surfaces of crystals or large molecules, the boundary of an autopoietic system is chemically distinct from the rest of the system, and it participates in metabolic processes by assembling itself and by selectively filtering incoming and outgoing molecules. /10

- The definition of a living system as an autopoietic network means that the phenomenon of life has to be understood as a property of the system as a whole. In the words of Pier Luigi Luisi, 'Life cannot be ascribed to any single molecular component (not even DNA or RNA!) but only to the entire bounded metabolic network. /10

- Autopoiesis provides a clear and powerful criterion for distinguishing between living and nonliving systems. For example, it tells us that viruses are not alive, because they lack their own metabolism. Outside living cells, viruses are inert molecular structures consisting of proteins and nucleic acids. A virus is essentially a chemical message that needs the metabolism of a living host cell to produce new virus particles, according to the instructions encoded in its DNA or RNA. The new particles are not built within the boundary of the virus itself, but outside of the host cell. /10

- Most people tend to believe that information about cellular processes is passed on to the next generation through the DNA when a cell divides and its DNA replicates. This is not at all what happens. When a cell reproduces, it passes on not only its genes, but also its membranes, enzymes, organelles—in short, the whole cellular network. The new cell is not produced from naked DNA, but from an unbroken continuation of the entire autopoietic network. Naked DNA is never passed on, because genes can only function when they are embedded in the epigenetic network. Thus life has unfolded for over three billion years in an uninterrupted process, without ever breaking the basic pattern of its self-generating networks. /12

- The starting point for this is the observation that all cellular structures exist far from equilibrium state—in other words, the cell would die—if the cellular metabolism did not use a continual flow of energy to restore structures as fast as they are decaying. This means that we need to describe the cell as an open system. Living systems are organizationally closed—they are autopoietic networks—but materially and energetically open. /13

- A dissipative structure, as described by Prigogine, is an open system that maintains itself in a state far from equilibrium, yet is nevertheless stable: the same overall structure is maintained in spite of an ongoing flow and change of components. Prigogine chose the term 'dissipative structures' to emphasize this close interplay between structure on the one hand and flow and change (or dissipation) on the other. /13

- The dynamics of these dissipative structures specifically include the spontaneous emergence of new forms of order. When the flow of energy increases, the system may encounter a point of instability, known as a 'bifurcation point,' at which it can branch off into an entirely new state where new structures and new forms of order may emerge. /13-14

- This spontaneous emergence of order at critical points of instability is one of the most important concepts of the new understanding of life. It is technically known as self-organization and is often referred to simply as 'emergence.' It has been recognized as the dynamic origin of development, learning and evolution. In other words, creativity—the generation of new forms—is a key property of all living systems. And since emergence is an integral part of the dynamics of open systems, we reach the important conclusion that open systems develop and evolve. Life constantly reaches out into novelty. /14

- The theory of dissipative structures, formulated in terms of non-linear dynamics, explains not only the spontaneous emergence of order, but also helps us to define complexity. Whereas traditionally the study of complexity has been a study of complex structures, the focus is now shifting from the structures to the processes of their emergence. For example, instead of defining the complexity of an organism in terms of the number of its different cell types, as biologists often do, we can define it as the number of bifurcations the embryo goes through in the organism's development. Accordingly, Brian Goodwin speaks of 'morphological complexity.' /14

- We have learned that a cell is a membrane-bounded, self-generating, organizationally closed metabolic network; that it is materially and energetically open, using a constant flow of matter and energy to produce, repair and perpetuate itself; and that it operates far from equilibrium, where new structures and new forms of order may spontaneously emerge, thus leading to development and evolution. /14

- Dissipative structures, then, are not necessarily living systems, but since emergence is an integral part of their dynamics, all dissipative structures have the potential to evolve. In other words, there is a 'prebiotic' evolution—an evolution of inanimate matter that must have begun some time before the emergence of living cells. This view is widely accepted among scientists today. /15

- The new thinking, as Morowitz emphasizes repeatedly, begins from the hypothesis that very early on, before the increase of molecular complexity, certain molecules assembled into primitive membranes that spontaneously formed closed bubbles, and that the evolution of molecular complexity took place inside these bubbles, rather than in a structureless chemical soup. /20

- The vesicle membranes are semipermeable, and thus various small molecules can enter the bubbles or be incorporated into the membrane. Among those will be chromophores, molecules that absorb sunlight. Their presence creates electric potentials across the membrane, and thus the vesicle becomes a device that converts light energy into electric potential energy. /20

- Eventually, a further refinement of this energy scenario takes place when the chemical reactions in the bubbles produce phosphates, which are very effective in the transformation and distribution of chemical energy. /21

- Cognition, according to Maturana and Varela, is the activity involved in the self-generation and self-perpetuation of living networks. In other words, cognition is the very process of life. The organizing activity of living systems, at all levels of life, is mental activity. The interactions of a living organism—plant, animal or human—with its environment are cognitive interactions. Thus life and cognition are inseparably connected. Mind—or, more accurately, mental activity—is immanent in matter at all levels of life. / 34

- In this new view, cognition involves the entire process of life —including perception, emotion, and behavior—and does not even necessarily require a brain and a nervous system. / 34

- According to the theory of autopoiesis, a living system couples to its environment structurally, i.e. through recurrent interactions, each of which triggers structural changes in the system. For example, a cell membrane continually incorporates substances from its environment into the cell's metabolic processes. An organism's nervous system changes its connectivity with every sense perception. These living systems are autonomous, however. The environment only triggers the structural changes; it does not specify or direct them. / 35

- As a living organism responds to environmental influences with structural changes, these changes will in turn alter its future behavior. In other words, a structurally coupled system is a learning system. / 35

- Continual structural changes in response to the environment —and consequently continuing adaptation, learning and development—are key characteristics of the behavior of all living beings. Because of its structural coupling, we can call the behavior of an animal intelligent but would not apply that term to the behavior of a rock. / 35

- It is interesting that the notion of consciousness as a process appeared in science as early as the late nineteenth century in the writings of William James, whom many consider the greatest American psychologist. James was a fervent critic of the reductionist and materialist theories that dominated psychology in his time, and an enthusiastic advocate of the interdependence of mind and body. He pointed out that consciousness is not a thing, but an ever-changing stream, and he emphasized the personal, continuous and highly integrated nature of this stream of consciousness. / 38

- The first, as mentioned above, is the recognition that consciousness is a cognitive process, emerging from complex neural activity. The second point is the distinction between two types of consciousness—in other words, two types of cognitive experiences—which emerge at different levels of neural complexity. /38

- The first type, known as 'primary consciousness,' arises when cognitive processes are accompanied by basic perceptual, sensory and emotional experience. Primary consciousness is probably experienced by most mammals and perhaps by some birds and other vertebrates. /39

- Reflective consciousness involves a level of cognitive abstraction that includes the ability to hold mental images, which allows us to formulate values, beliefs, goals and strategies. This evolutionary stage is of central relevance to the main theme of this book—the extension of the new understanding of life to the social domain—because with the / evolution of language arose not only the inner world of concepts and ideas, but also the social world of organized relationships and culture. /40

- When carbon, oxygen, and hydrogen atoms bond in a certain way to form sugar, the resulting compound has a sweet taste. The sweetness resides neither in the C, nor in the O, nor in the H; it resides in the pattern that emerges from their interaction. It is an emergent property. Moreover, strictly speaking, the sweetness is not a property of the chemical bonds. It is a sensory experience that arises when the sugar / molecules interact with the chemistry of our taste buds, which in turn causes a set of neurons to fire in a certain way. The experience of sweetness emerges from that neural activity. /41-42

- Phenomenology is an important branch of modern philosophy, founded by Edmund Husserl at the beginning of the twentieth century and developed further by many European philosophers, including Martin Heidegger and Maurice Merleau-Ponty. The central concern of phenomenology is the disciplined examination of experience, and the hope of Husserl and his followers was, and is, that a true science of experience would eventually be established in partnership with the natural sciences. /45

- The specific neural mechanism proposed by Varela for the emergence of transitory experiential states is a resonance phenomenon known as 'phase-locking', in which different brain regions are interconnected in such a way that their neurons fire in synchrony. Through this synchronization of neural activity, temporary 'cell assemblies' are formed, which may consist of widely dispersed neural circuits. /50

- Humberto Maturana was one of the first scientists to link the biology of human consciousness to language in a systematic way. He did so by approaching language through a careful analysis of communication within the framework of the Santiago Theory of Cognition. Communication, according to Maturana, is not the transmission of information but rather the coordination of behavior between living organisms through mutual structural coupling. In these recurrent interactions, the living organisms change together through their mutual triggering of structural changes. Such mutual coordination is the key characteristic of communication for all living organisms, with or without nervous systems, and it becomes more and more subtle and elaborate with nervous systems of increasing complexity. /53

- Language arises when a level of abstraction is reached at which there is communication about communication. /53

- How living organisms categorize depends on their sensory apparatus and their motor systems; in other words, it depends on how they are embodied. This is true not only for animals, plants, and microorganisms, but also for human beings, as cognitive scientists have recently discovered. Although some of our categories are the result of conscious reasoning, most of them are formed automatically, and unconsciously as a result of the specific nature of our bodies and brains. /62

- All these manifestations are part of the process of cognition, and at each new level they involve corresponding neural and bodily structures. As the recent discoveries in cognitive linguistics have shown, the human mind, even in its most abstract manifestations, is not separate from the body but arises from it and is shaped by it. /65

- The unified, post-Cartesian view of mind, matter, and life also implies a radical reassessment of the relationships between humans and animals. Throughout most of Western philosophy, the capacity to reason was seen as a uniquely human characteristic, distinguishing us from all other animals. The communication studies with chimpanzees / have exposed the fallacy of this belief in the most dramatic of ways. They make it clear that the cognitive and emotional lives of animals and humans differ only by degree; that life is a great continuum in which differences between species are gradual and evolutionary. /65-66

▸ Spirituality, then, is always embodied. /68

▸ When we study the living systems from the perspective of form, we find that their pattern of organization is that of a self-generating network. From the perspective of matter, the material structure of a living system is a dissipative structure, i.e. an open system operating far from equilibrium. From the process perspective, finally, living systems are cognitive systems in which the process of cognition is closely linked to the pattern of autopoiesis. In a nutshell, this is my synthesis of the new scientific understanding of life. /71

▸ Of course, no scientist would deny the existence of patterns and processes, but most of them think of a pattern as an emergent property of matter, an idea abstracted from matter, rather than a generative force. /72

▸ To focus on material structures and the forces between them, and to view the patterns of organization resulting from these forces as secondary emergent phenomena has been very effective in physics and chemistry, but when we come to living systems this approach is no longer adequate. /72

▸ Such a systemic understanding is based on the assumption that there is a fundamental unity to life, that different living systems exhibit similar patterns of organization. This assumption is supported by the observation that evolution has proceeded for billions of years by using the same patterns again and again. As life evolves, these patterns tend to become more and more elaborate, but they are always variations on the same basic themes. /81

▸ The network, in particular, is one of the very basic patterns of organization in all living systems. At all levels of life—

from the metabolic networks of cells to the food webs of ecosystems—the components and processes of living systems are interlinked in network fashion. Extending the systemic understanding of life to the social domain, therefore, means applying our knowledge of life's basic patterns and principles of organization, and specifically our understanding of living networks, to social reality. / 81

- Through this shared context of meaning individuals acquire identities as members of the social network, and in this way the network generates its own boundary. It is not a physical boundary but a boundary of expectations, of confidentiality and loyalty, which is continually maintained and renegotiated by the network itself. / 83

- To understand the meaning of anything we need to relate it to other things in its environment, in its past, or in its future. Nothing is meaningful in itself. / 84

- Meaning is essential to human beings. We continually need to make / sense of our outer and inner worlds, find meaning in our environment and in our relationships with other humans, and act according to that meaning. This includes in particular our need to act with a purpose or goal in mind. Because of our ability to project mental images into the future we act with our conviction, valid or invalid, that our actions are voluntary, intentional, and purposeful. / 85

- In his classic text, *Culture*, historian Raymond Williams traces the meaning of the word back to its early use as a noun denoting a process: the culture (i.e. cultivation) of crops, or the culture (i.e. rearing and breeding) of animals. In the seventeenth century this meaning was extended metaphorically to the active cultivation of the human mind; and in the late eighteenth century, when the word was borrowed from the French by German writers (who first spelled it *Cultur* and subsequently *Kultur*), it acquired the meaning of a distinctive way of life of a people. In the nineteenth century the plural 'cultures' became especially important in the development of comparative anthropology, where it has continued to designate distinctive ways of life. / 86

- In the meantime, the older use of 'culture' as the active cultivation of the mind continued. Indeed, it expanded and diversified, covering a range of meanings from a developed state of mind ('a cultured person') to the process of this development ('cultural activities') to the means of these processes (administered, for example, by a 'Ministry of Culture'). /86

- Over the past ten years, I have been invited to speak at quite a few business conferences, and at first I was very puzzled when I encountered the strongly felt need for organizational change. Corporations seemed to be more powerful than ever; business was clearly dominating politics; and the profits and shareholder values of most companies were rising to unprecedented heights. Things seemed to be going very well indeed for business, so why was there so much talk about fundamental change? /97

- [I]t is becoming more and more apparent that our complex industrial systems, both organizational and technological, are the main driving force of global environmental / destruction, and the main threat to the long-term survival of humanity. To build a sustainable society for our children and future generations, we need to fundamentally redesign many of our technologies and social institutions so as to bridge the wide gap between human design and the ecologically sustainable systems of nature. /98-99

- Organizations need to undergo fundamental changes, both in order to adapt to the new business environment and to become ecologically sustainable. This double challenge is urgent and real, and the recent extensive discussions of organizational change are fully justified. However, despite these discussions and some anecdotal evidence of successful attempts to transform organizations, the overall track record is very poor. /99

- It is common to hear that people in organizations resist change. In / reality, people do not resist change; they resist having change imposed on them. Being alive, individuals and their communities are both stable and subject to change and development, but their natural change processes are very different from the organizational changes designed by 'reengineering' experts and mandated from the top. /99-100

- To run properly, a machine must be controlled by its operators, so that it will function according to their instructions. Accordingly, the whole thrust of classical management theory is to achieve efficient operations through top-down control. Living beings, on the other hand, act autonomously. They can never be controlled like machines. To try and do so is to deprive them of their aliveness. / 104

- The need to have all changes designed by management and imposed upon the organization tends to generate bureaucratic rigidity. There is no room for flexible adaptations, learning, and evolution in the machine metaphor, and it is clear that organizations managed in strictly mechanistic ways cannot survive in today's complex, knowledge-oriented and rapidly changing business environment. / 105

- He [de Geus] identifies to sets of characteristics. One is a strong sense of community and collective identity around a set of common values; a community in which all members know that they will be supported in their endeavors to achieve their own goals. The other set of characteristics is openness to the outside world, tolerance for the entry of new individuals and ideas, and consequently a manifest ability to learn and adapt to new circumstances. / 105

- These considerations imply that the most effective way to enhance an organization's potential for creativity and learning, to keep it vibrant and alive, is to support and strengthen its communities of practice. The first step in this endeavor will be to provide the social space for information communications to flourish. Some companies may create special coffee counters to encourage informal gatherings; others may use bulletin boards, the company newsletter, a special library, offsite retreats or online chat rooms for the same purpose. If widely publicized within the company so that support by management is evident, these measures will liberate people's energies, stimulate creativity, and set processes of change in motion. / 111

- We are dealing here with a crucial difference between a living system and a machine. A machine can be controlled; a living system, according to the systemic understanding of life, can only be disturbed. In other words, organizations cannot be controlled through direct interventions, but they can be influenced by giving impulses rather than instructions. / 112

- There is no need to push, pull, or bully it to make it change. Force or energy are not the issue; the issue is meaning. Meaningful disturbances will get the organization's attention and will trigger structural changes. /112

- In terms of our previous discussion of power, we could say that the shift from domination to partnership corresponds to a shift from coercive power, which uses threats or sanctions to assure adherence to orders, to compensatory power, which tries to make instructions meaningful through persuasion and education. /114

- Whereas explicit knowledge can be communicated and documented through language, tacit knowledge is acquired through experience and often remains intangible. /115

- Tacit knowledge is created by the dynamics of culture resulting from a network of (verbal and nonverbal) communications within a community of practice. Organizational learning, therefore, is a social phenomenon, because the tacit knowledge on which all explicit knowledge is based is generated collectively. /115

- The traditional idea of a leader is that of a person who is able to hold a vision, to articulate it clearly and to communicate it with passion and charisma. /121

- The ability to hold a clear vision of an ideal form, or state of affairs, is something that traditional leaders have in common with designers. /122

- Holding a vision is central to the success of any organization, because all human beings need to feel that their actions are meaningful and geared toward specific goals. At all levels of the organization, people need to have a sense of where they are going. A vision is a mental image of what we want to achieve, but visions are much more complex than concrete goals and tend to defy expression in ordinary, rational terms. Goals can be measured, while vision is qualitative and much more intangible. /122

- Whenever we need to express complex and subtle messages, we make use of metaphors, and thus it is not surprising that metaphors play a crucial role in formulating an organization's vision. Often, the vision remains unclear as long as we try to explain it, but suddenly comes into focus when we find the right metaphor. The ability to express a vision in metaphors, to articulate it in such a way that it is understood and embraced by all, is an essential quality of leadership. /122

- The experience of the critical instability that precedes the emergence of novelty may involve uncertainty, fear, confusion, or self-doubt. Experienced leaders recognize these emotions as integral parts of the whole dynamic and create a climate of trust and mutual support. In / today's turbulent global economy this is especially important, because people are often in fear of losing their jobs as a consequence of corporate mergers or other radical structural changes. This fear generates a strong resistance to change, hence building trust is essential. /123-124

- The problem is that people at all levels want to be told what concrete results they can expect from the change process, while managers themselves do not know what will emerge. During this chaotic phase, many managers tend to hold things back rather than communicating honestly and openly, which means that rumors fly and nobody knows what information to trust. /124

- During the change process some of the old structures may fall apart, but if the supportive climate and the feedback loops in the network of communications persist, new and more meaningful structures are likely to emerge. When that happens, people often feel a sense of wonder and elation, and now the leader's role is to acknowledge these emotions and provide opportunities for celebration. /124

- Finally, leaders need to be able to recognize emergent novelty, articulate it and incorporate it into the organization's design. Not all emergent solutions will be viable, however, and hence a culture fostering emergence must include the freedom to make mistakes. In such a culture, experimentation is encouraged and learning is valued as much as success. /124

- The more we understand the nature of life and become aware of how alive an organization can be, the more painfully we notice the life-draining nature of our current economic system. /126

- These economic pressures are applied with the help of ever more sophisticated information and communication technologies, which have created a profound conflict between biological time and computer time. New knowledge arises, as we have seen, from chaotic processes of emergence that take time. Being creative means being able to relax into uncertainty and confusion. In most organizations this is becoming increasingly difficult, because things move far too fast. People feel that they have hardly any time for quiet reflection, and since reflective consciousness is one of the defining characteristics of human nature, the results are profoundly dehumanizing. /126

- This new economy is structured around flows of information, power, and wealth in global financial networks that rely decisively on advanced information and communication technologies. It is shaped in very fundamental ways by machines, and the resulting economic, social, and cultural environment is not life-enhancing but life-degrading. It has triggered a great deal of resistance, which may well coalesce into a worldwide movement to change the current economic system by organizing its financial flows according to a different set of values and beliefs. The systemic understanding of life makes it clear that in the coming years such a change will be imperative not only for the well-being of human organizations, but also for the survival and sustainability of humanity as a whole. /128

- With the creation of the World Trade Organization (WTO) in the mid-1990s, economic globalization, characterized by 'free trade' was hailed by corporate leaders and politicians as a new order that would benefit all nations, producing worldwide economic expansion whose wealth would trickle down to all. However, it soon became apparent to increasing numbers of environmentalists and grassroots activists that the new economic rules established by the WTO were manifestly unsustainable and were producing a multitude of interconnected fatal consequences—social disintegration, a breakdown of democracy, more rapid and extensive deterioration of the environment, the spread of new diseases, and increasing poverty and alienation. /129

- Global currency markets alone involve the daily exchange of over two trillion dollars, and since these markets largely determine the value of any national currency, they contribute significantly to the inability of governments to control economic policy. /139

- The recent crashes of the financial markets threw approximately 40 percent of the world's population into deep recession. /140

- The logic of this automaton is not that of traditional market rules, and the dynamics of the financial flows it sets in motion is currently beyond the control of governments, corporations, and financial institutions, regardless of their wealth and power. However, because of the great versatility and accuracy of the new information and communication technologies, effective regulation of the global economy is technically feasible. The critical issue is not technology, but politics and human values. /141

- According to the United Nation's Human Development Report, the difference in per capita income between the North and South tripled from $5,700 in 1960 to $15,000 in 1993. The richest 20 percent of the world's people now own 85 percent of its wealth, while the poorest 20 percent (who account for 80 percent of the total world population) owns just 1.4 percent. The assets of the three richest people in the world alone exceed the combined GNP of all least developed countries and their 600 million people. /144

- The increase of poverty and especially of extreme poverty, seems to be a worldwide phenomenon. Even in the United States, 15 percent of the population (including 25 percent of all children) now lives below the poverty line. One of the most striking features of the 'new poverty' is homelessness, which skyrocketed in American cities during the 1980's and remains at high levels today. /144

- The Fourth World is populated by millions of homeless, impoverished, and often illiterate people who move in and out of paid work, many of them drifting into the criminal economy. They experience multiple crises in their lives, including hunger, disease, drug addiction, and imprisonment—the ultimate form of social exclusion. Once their poverty turns into misery, they may easily find themselves caught in a downward spiral of marginality from which it is almost impossible to escape. /145

- Global capitalism does not alleviate poverty and social exclusion; on the contrary, it exacerbates them. The Washington consensus has been blind to this effect because corporate economists have traditionally excluded the social costs of economic activity from their models. Similarly, most conventional economists have ignored the new economy's environmental cost—the increase and acceleration of global environmental destruction, which is as severe, if not more so, than its social impact. /146

- In Taiwan, agricultural and industrial poisons have severely polluted nearly every major river. In some places, the water is not only devoid of fish and unfit to drink, but is actually combustible. The level of air pollution is twice that considered harmful in the United States; cancer rates have doubled since 1965, and the country has the world's highest incidence of hepatitis. In principle, Taiwan could use its new wealth to clean up its environment, but competitiveness in the global economy is so extreme that environmental regulations are eliminated rather than strengthened in order to lower the costs of industrial production. /147

- One of the tenets of neoliberalism is that poor countries should concentrate on producing a few special goods for export in order to obtain foreign exchange, and should import most other commodities. This emphasis has led to the rapid depletion of the natural resources required to produce export crops in country after country—diversion of fresh water from vital rice paddies to prawn farms; a focus on water-intensive crops, such as sugar cane, that result in dried-up riverbeds; conversion of good agricultural land into cash-crop plantations; and forced migration of large numbers of farmers from their lands. All over the world there are countless examples of how economic globalization is worsening environmental destruction. /147

- Studies in Germany have shown that the contribution of nonlocal food production to global warming is between six and twelve times higher than that of local production, due to increased CO_2 emissions. /147

- The destruction of the natural environment in Third World countries goes hand in hand with the dismantling of rural people's traditional, largely self-sufficient ways of life, as American television programs and transnational advertising agencies promote glittering images of modernity to billions of people all over the globe without mentioning that the lifestyle of endless material consumption is utterly unsustainable. /148

- Edward Goldsmith estimates that, if all Third World countries were to reach the consumption level of the United States by the year 2060, the annual environmental damage from the resulting economic activities would be 220 times what it is today, which is not even remotely conceivable. /148

- Nevertheless, it would be false to think that a few megacorporations control the world. To begin with, real economic power has shifted to the global financial networks. Every corporation depends on what happens in those complex networks, which nobody controls. There are thousands of corporations today, all of whom compete and cooperate at the same time, and no individual corporation can dictate conditions. /152

- It is instructive to compare this situation with ecological networks. Although it may seem that in an ecosystem some species are more powerful than others, the concept of power is not appropriate, because nonhuman species (with the exception of some primates) do not force individuals to act in accordance with preconceived goals. There is dominance, but it is always acted out within a larger context of cooperation, even in predator-prey relationships. The manifold species in an ecosystem do not form hierarchies, as is often erroneously stated, but exist in networks nested within networks. /152

- In spite of the constant barrage of advertising and the billions of dollars spent on it every year, studies have shown repeatedly that media advertising has virtually no specific impact on consumer behavior. This startling discovery is further evidence for the observation that human beings, like all living systems, cannot be directed but can only be disturbed. As we have seen, choosing what to notice and how to respond is the very essence of being alive. /154

- Geneticists soon discovered that there is a huge gap between the ability to identify genes that are involved in the development of disease and the understanding of their precise function, let alone their manipulation to obtain a desired outcome. As we now know, this gap is a direct consequence of the mismatch between the linear causal chains of genetic determination and the nonlinear epigenetic networks of biological reality. /178

- The reality of genetic engineering is much more messy. At the current state of the art, geneticists cannot control what happens in the organism. They can insert a gene into the nucleus of a cell with the help of a specific gene transfer vector, but they never know whether the cell will incorporate it into its DNA, nor where the new gene will be located, nor what effects this will have on the organism. Thus, genetic engineering proceeds by trial and error in a way that is extremely wasteful. The average success rate of genetic experiments is only about 1 percent, because the living background of the host organism, which determines the outcome of the experiment, remains largely inaccessible to the engineering mentality that underlies our current biotechnologies. /178

- The real ethical problems surrounding the current cloning procedure are rooted in the biological developmental problems it generates. They are a consequence of the crucial fact that the manipulated cell from which the embryo grows is a hybrid of cellular components from two different animals. Its nucleus stems from one organism, while the rest of the cell, which contains the entire epigenetic network, stems from another. Because of the enormous complexity of the epigenetic network and its interactions with the genome, the two components will only very rarely be compatible. /183

- [1] With the new chemicals, farming became mechanized and energy intensive, favoring large corporate farmers with sufficient capital, and forcing most of the traditional single-family farmers to abandon their land. All over the world, large numbers of people have left rural areas and joined the masses of urban unemployed as victims of the Green Revolution. /186

- [2] The long-term effects of excessive chemical farming have been disastrous for the health of the soil and for human health, for our social relations, and for the entire natural environment on which our well-being and future survival depends. As the same crops were planted and fertilized synthetically year after year, the balance of the ecological processes in the soil was disrupted; the amount of organic matter diminished, and with it the soil's ability to retain moisture. The resulting changes in soil texture entailed a multitude of interrelated harmful consequences—loss of humus, dry and sterile soil, wind and water erosion, and so on. /186

- [3] The ecological imbalance caused by monocultures and excessive use of chemicals also resulted in enormous increases in pests and crop diseases, which farmers countered by spraying even larger doses of pesticides in vicious cycles of depletion and destruction. The hazards for human health increased accordingly as more and more toxic chemicals seeped through the soil, contaminated the water table and showed up in our food. /186-187

- Through a series of massive mergers and because of the tight control afforded by genetic technologies, an unprecedented concentration of ownership and control over food production is now under way. The top ten agrochemical companies control 85 percent of the global market; the top five control virtually the entire market for genetically modified (GM) seeds. Monsanto alone bought into the major seed companies in India and Brazil, in addition to buying numerous biotech companies, while DuPont bought Pioneer Hi-Bred, the world's largest seed company. The goal of these corporate giants is to create a single world agricultural system in which they would be able to control all stages of food production and manipulate both food supplies and / prices. /187-188

- Biotechnology proponents have argued repeatedly that GM seeds are crucial to feed the world, using the same flawed reasoning that was advanced for decades by the proponents of the Green Revolution. Conventional food production, they maintain, will not keep pace with the world population. /188

- Development agencies have known for a long time that there is a direct relationship between the prevalence of hunger and a country's population density or growth. There is widespread hunger in densely populated countries like

Bangladesh and Haiti, but also in sparsely populated ones like Brazil and Indonesia. Even in the United States, in the midst of super-abundance, there are between 20 and 30 million malnourished people. /188

- The root causes of hunger around the world are unrelated to food production. They are poverty, inequality, and lack of access to food and land. /189

- Recent experimental trials have shown that GM seeds do not increase crop yields significantly. Moreover, there are strong indications that the widespread use of GM crops will not only fail to solve the problem of hunger but, on the contrary, may perpetuate and even aggravate it. If transgenic seeds continue to be developed and promoted exclusively by private corporations, poor farmers will not be able to afford them, and if the biotech industry continues to protect its / products by means of patents that prevent farmers from storing and trading seeds, the poor will become further dependent and marginalized. According to a recent report by the charitable organization Christian Aid, 'GM crops are ... creating classic preconditions for hunger and famine. Ownership of resources concentrated in too few hands—inherent in farming based on patented proprietary products—and a food supply based on too few varieties of crops widely planted are the worst option for food security. /190

- Organic farming preserves and sustains the great ecological cycles, integrating their biological processes into the processes of food production. When soil is cultivated organically, its carbon content increases, and thus organic farming contributes to reducing global warming. Physicist Amory Lovins estimates that increasing the carbon content of the world's depleted soils at plausible rates would absorb about as much carbon as all human activity emits. /191

- [1] Scientists at a recent international conference on sustainable agriculture in Bellagio, Italy, reported that a series of large-scale experimental projects around the world that tested agroecological techniques—crop rotation, intercropping, use of mulches and compost, terracing, water harvesting, etc.—yielded spectacular results. Many / of these were achieved in resource-poor areas that had been deemed incapable of producing food supplies. For example, agroecological projects involving about 730,000 farm households across Africa resulted in yield increases of between 80 and 100 percent, while decreasing production costs, increasing cash incomes of households dramatically—sometimes by as much as ten times. Again and again it was demonstrated that organic farming not only increased production and offered a wide range of ecological benefits, but also empowered the farmers. As one Zambian farmer put it, 'Agroforestry has restored my dignity. My family is no longer hungry; I can even help my neighbors now.' /191-192

- [2] In southern Brazil, the use of cover crops to increase soil activity and water retention enabled 400,000 farmers to increase maize and soybean yields by over 60 percent. In the Andean region, increases in crop varieties resulted in twentyfold increases in yields and more. In Bangladesh, an integrated rice-fish program raised rice yields by 8 percent and farmers' incomes by 50 percent. In Sri Lanka, integrated pest and crop management increased rice yields by 11 to 44 percent while augmenting net incomes by 38 to 178 percent. /192

- The risks of current biotechnologies in agriculture are a direct consequence of our poor understanding of genetic function. We have only recently come to realize that all biological processes involving genes are regulated by the cellular networks in which genomes are imbedded, and that the patterns of genetic activity change continually in response to changes in the cellular environment. Biologists are only just beginning to shift their attention from genetic structures to metabolic networks, and they still know very little about the complex dynamics of these networks. /193

- We also know that all plants are embedded in complex ecosystems above the ground and in the soil, in which inorganic and organic matter moves in continual cycles. Again, we know very little about these ecological cycles and networks—partly because for many decades the dominant genetic determinism resulted in a severe distortion of biological research, with most of the funding going into molecular biology and very little into ecology. /193

- Since the cells and regulatory networks of plants are relatively simpler than those of animals, it is much easier for geneticists to insert foreign genes into plants. The problem is that once the foreign gene is in the plant's DNA and the resulting transgenic crop has been planted, it becomes part of an entire ecosystem. The scientists working for biotech companies know very little about the ensuing biological processes, and even less about the ecological consequences of their actions. /193

- The most widespread use of plant biotechnology has been to develop herbicide-tolerant crops in order to boast the sales of particular herbicides. There is a strong likelihood that the transgenic plants will cross-pollinate with wild relatives in their surroundings, thus creating herbicide-resistant 'superweeds.' Evidence indicates that such gene flows between transgenic crops and wild relatives are already occurring. /193

- To defend their practices, biotech supporters often claim that genetic engineering is like conventional breeding—a continuation of the age-old tradition of shuffling genes to obtain superior crops and live-stock. Sometimes they even argue that our modern biotechnologies represent the latest stage in nature's adventure of evolution. Nothing could be farther from the truth. To begin with, the pace of gene alteration through biotechnology is several orders of magnitude faster than nature's. No ordinary plant breeder would be able to alter the genomes of half of the world's soybeans in just three years. Genetic modification of crops is undertaken with incredible haste, and transgenic crops are planted massively without proper testing of the short- and long-term impacts on ecosystems and human health. These untested and potentially hazardous GM crops are now spreading all over the world, creating irreversible risks. /194

- Most of the ecological hazards associated with herbicide-resistant crops, such as Monsanto's Roundup Ready soybeans, derive from the ever-increasing use of the company's herbicide. Since resistance to that specific herbicide is the crop's only—and widely advertised—benefit, farmers are naturally led to use massive amounts of the weed-killer. It is well documented that such massive use of a single chemical greatly boosts herbicide resistance in weed populations, which triggers a vicious cycle of more and more intensive spraying. Such use of toxic chemicals in agriculture is especially harmful to consumers. When plants are sprayed repeatedly with a weed-killer, they retain chemical residues that show up in our food. Moreover, plants grown in the presence of massive amounts of herbicides can suffer from stress and will typically respond by over- or underproducing certain substances.

Herbicide-resistant members of the bean family are known to produce higher levels of plant oestrogens, which may cause severe dysfunctions in human reproductive systems, especially in boys. /196

- [1] Monsanto is now facing an increasing number of lawsuits from farmers who had to cope with (...) unexpected side effects. For example, the balls of their GM cotton were deformed and dropped off in thousands of acres in the Mississippi Delta; their GM canola seeds had to be pulled off the Canadian market because of contaminations with a hazardous gene. Similarly, Calgene's Flavr-Savr tomato, engineered for improved shell life, was a commercial disaster and soon disappeared. Transgenic potatoes intended for human consumption caused a series of serious health problems when they were fed to rats, including tumor growth, liver atrophy, and shrinkage of the brain. /198

- In the animal kingdom, where cellular complexity is much higher, the side effects in genetically modified species are much worse. 'Super-salmon' which were engineered to grow as fast as possible, ended up with monstrous heads and died from not being able to breathe or feed properly. Similarly, a 'superpig' with a human gene for a growth hormone turned out ulcerous, blind, and impotent. /198

- The most horrifying and by now best-known story is probably that of the genetically altered hormone called 'recombinant bovine growth hormone,' which has been used to stimulate milk production in cows despite the fact that American dairy farmers have produced vastly more milk than people can consume for the past fifty years. The effects of this genetic engineering folly on the cow's health are serious. They include bloat, diarrhea, diseases of the knees and feet, cystic ovaries, and many more. Besides, their milk may contain a substance that has been implicated in human breast and stomach cancers. /198

- In the United States, the biotech industry has persuaded the Food and Drug Administration (FDA) to treat GM food as 'substantially equivalent' to traditional food, which allows food producers to evade normal testing by the FDA and the Environmental Protection Agency (EPA), and also leaves it to the companies' own discretion as to whether to label their products as genetically modified. Thus, the public is kept unaware of the rapid spread of transgenic foods and scientists will find it much harder to trace harmful effects. Indeed, buying organic is now the only way to avoid GM foods. /199

- Ciba-Geigy merged with Sandoz to become Novartis; Hoechst and Rhone Poulenc became Aventis; and Monsanto now owns and controls several large seed companies. /199

- As Vandana Shiva reminds us, the Latin root of the word 'resource' is *resurgere* ('to rise again'). In the ancient meaning of the term, a natural resource, like all of life, is inherently self-renewing. This profound understanding of life is denied by the new life sciences corporations when they prevent life's self-renewal in order to turn natural resources into profitable raw materials for industry. They do so through a combination of genetic alterations (including the terminator technologies) and patents, which do violence to time-honored farming practices that respect the cycles of life. /200

- Since a patent is traditionally understood as the exclusive right to the use and selling of an invention, it seems strange that biotech companies today are able to patent living organisms, from bacteria to human cells. The history of this achievement is an amazing story of scientific and legal sleight of hand. The patenting of life-forms became common practice in the 1960's when property rights were given to plant breeders for new varieties of flowers obtained through human intervention and ingenuity. It took the international legal community less than twenty years to move from this seemingly less harmless patenting of flowers to the monopolization of life. /200-201

- Indeed, in 1980 the U.S. Supreme Court handed down the landmark decision that genetically modified microorganisms could be patented. /201

- The patents now granted to biotech companies (…) cover not only the methods by which DNA sequences are isolated, identified, and transferred, but also the underlying genetic material itself. Moreover, existing national laws and international conventions that specifically prohibit the patenting of essential natural resources, such as food and plant-derived medicine, are now being altered in accordance with the corporate view of life as a profitable commodity. /201

- These exploitative practices are legalized by the WTO's narrow definition of intellectual property rights (IPRs), which recognizes knowledge as patentable only when it is expressed within the framework of Western science./201

- The development of such new biotechnologies will be a tremendous intellectual challenge, because we still do not understand how nature developed 'technologies' during billions of years of evolution that are far superior to our human designs. How do mussels produce glue that sticks to anything in water? How do spiders spin a silk thread that, ounce for ounce, is five times stronger than steel? How do abalone grow a shell that is twice as tough as our high-tech ceramics? How do these creatures manufacture their miracle materials in water, at room temperature, silently, and without any toxic byproducts? /204

- Scientists at the University of Washington have studied the molecular structure and assembly process of the smooth inner coating of abalone shells, which shows delicate swirling color patterns and is hard as nails. They were able to mimic the assembly process at ambient temperatures and create a hard, transparent material that could be an ideal coating for the windshields of ultralight electric cars. German researchers have mimicked the bumpy, self-cleaning micro-surface of the lotus leaf to produce a paint that will do the same for buildings. Marine biologists and biochemists have spent many years analyzing the unique chemistry used by blue mussels to secrete an adhesive that bonds underwater. They are now exploring potential medical applications that would allow surgeons to create bonds between ligaments and tissues in a fluid environment. Physicists have teamed up with biochemists in several laboratories to examine the complex structures and processes of photosynthesis, eventually hoping to mimic them in new kinds of solar cells. /204-205

- [1] While these exciting developments are taking place, however, the central assertion of genetic determinism that genes determine behavior is still perpetuated by many geneticists, in biotechnology companies as well as in the academic world. One has to wonder whether these scientists really believe that our behavior is determined by our genes, and if not, why they keep up this façade. /205

- [2] Discussions of this issue with molecular biologists have shown me that there are several reasons why scientists feel that they have to perpetuate the dogma of genetic determinism in spite of mounting contrary evidence. Industrial scientists are often hired for specific, narrowly defined projects, work under strict supervision, and are forbidden to discuss the broader implications of their research. They are required to sign confidentiality clauses to that effect. In biotechnology companies, in particular, the pressure to conform with the official doctrine of genetic determinism is enormous. /205

- [3] In the academic world the pressures are different but, unfortunately, almost equally strong. Because of the tremendous cost of genetic research, biology departments increasingly form partnerships with biotechnology corporations to receive substantial grants that shape the nature and direction of their research. /205

- [4] Biologists are used to formulating their grant proposals in terms of genetic determinism, because they know that this is what gets funded. They promise their funders that certain results will be derived from the future knowledge of genetic structure even though they know well that scientific advances are always unexpected and unpredictable. They learn to adopt this double standard during the years as graduate students and then keep it up throughout their academic careers. /205-206

- The new global capitalism has also created a global criminal economy that profoundly affects national and international economies and politics; it has threatened and destroyed local communities around the world; and with the pursuit of an ill-conceived biotechnology it has invaded the sanctity of life by attempting to turn diversity into monoculture, ecology into engineering, and life itself into a commodity. /207

- In late 2000, the authoritative Intergovernmental Panel on Climatic Change (IPCC) published its strongest consensus statement to date that human release of carbon dioxide and other greenhouse gases 'contributed significantly to the observed warming over the last fifty years.' /208

- Countries around the world with vastly different cultural traditions are increasingly homogenized through relentless proliferation of the same restaurant franchises, hotel chains, high-rise architecture, superstores, and shopping malls. The result, in Vandana Shiva's apt phrase, is an increasing 'monoculture of the mind.' /213

- At the 2001 meeting of the World Economic Forum in Davos, the exclusive club of representatives from big business, some of the leading players admitted for the first time that globalization has no future unless it is designed to be inclusive, ecologically sustainable, and respectful of human rights and values. /214

- As members of the human community, our behavior should reflect a respect of human dignity and basic human rights. Since human life encompasses biological, cognitive, and social dimensions, human rights should be respected in all three of these dimensions. The biological dimension includes the right to a healthy environment and to secure and healthy food; honoring the integrity of life also includes the rejection of the patenting of life-forms. Human rights in the cognitive dimension include the right of access to education and knowledge, as well as the freedom of opinion and expression. In the social dimension, finally, the first human right—in the words of the UN Declaration of Human Rights—is 'the right to life, liberty, and security of person.' There is a wide range of human rights in the social dimension—from social justice to the right of peaceful assembly, cultural integrity, and self-determination. /215

- [T]he rise of the network society has gone hand in hand with the decline of the sovereignty, authority, and legitimacy of the nation-state. At the same time, mainstream religions have not developed an ethic appropriate for the age of globalization, while the legitimacy of the traditional patriarchal family is being challenged by profound redefinitions of gender relationships, family, and sexuality—the main institutions of traditional civil society are breaking down. /219

- The governments of France, Italy, Greece, and Denmark announced that they would block the approval of new GM crops in the European Union. The European Commission made the labeling of GM foods mandatory, as did the governments of Japan, South Korea, Australia, and Mexico.

In January 2000, 130 nations signed the groundbreaking Cartagena Protocol on Biosafety in Montreal, which gives nations the right to refuse entry to any genetically modified forms of life, despite vehement opposition from the United States. /228

- The concept of sustainability was introduced in the early 1980s by Lester Brown, founder of the Worldwatch Institute, who defined a sustainable society as one that is able to satisfy its needs without diminishing the chances of future generations. /229

- The key to an operational definition of ecological sustainability is the realization that we do not need to invent sustainable human communities from scratch but can model them after nature's ecosystems, which are sustainable communities of plants, animals, and microorganisms. Since the outstanding characteristic of the Earth household is its inherent ability to sustain life, a sustainable human community is one designed in such a manner that its ways of life, businesses, economy, physical structures, and technologies do not interfere with nature's inherent ability to sustain life. /230

- Ecodesign is a process in which our human purposes are carefully meshed with the larger patterns and flows of the natural world. Ecodesign principles reflect the principles of organization that nature has evolved to sustain the web of life. To practice industrial design in such a context requires a fundamental shift in our attitude toward nature./233

- The principle 'waste equals food' means that all products and materials manufactured by industry, as well as the wastes generated in the manufacturing process, must eventually provide nourishment for something new. /234

- Such ecological clusters of industries have actually been initiated in many parts of the world by an organization called Zero Emissions Research and Incentives (ZERI), founded by business entrepreneur Gunter Pauli in the early 1990s. Pauli introduced the notion of industrial clustering by promoting the principle of zero emissions and making it the very core of the ZERI concept. /234

- To appreciate how radical an approach this is, we need to realize that our current businesses throw away most of the resources they take from nature. For example, when we extract cellulose from wood to make paper, we cut down forests but use only 20 to 25 percent of the trees, discarding the remaining 75 to 80 percent as waste. Beer breweries extract only 8 percent of the nutrients from barley or rice for fermentation; palm oil is a mere 4 percent of the palm tree's biomass; and coffee beans are 3.7 percent of the coffee bush. /234

- The clustering of these productive systems inexpensively generates several revenue streams in addition to the original coffee beans—from poultry, mushrooms, vegetables, beef, and pork—while creating jobs in the local community. The results are beneficial both to the environment and the community; there are no high investments; and there is no need for the coffee farmers to give up their traditional livelihood. /236

- The places of production are usually close to those of consumption, which eliminates or radically reduces transportation costs. No single production unit tries to maximize its output, because this would only unbalance the system. Instead, the goal is to optimize the production processes of each component, while maximizing the productivity and ecological sustainability of the whole. /237

- Similar agricultural clusters, with beer breweries as their center instead of coffee farms, are operating in Africa, Europe, Japan, and other parts of the world. Other clusters have aquatic components; for example, a cluster in southern Brazil includes the farming of highly nutritious spirulina algae in the irrigation channels of the rice fields (which otherwise are used only once a year). The spirulina is used as special enrichment in a 'ginger cookie' program in rural schools to fight widespread malnutrition. This generates additional revenue for the rice farmers while responding to a pressing social need. /237

- The combined value created by the whole is always greater than the sum of the values that would be generated by independently operating components. /239

- In a sustainable industrial society, all products, materials, and wastes will either be biological or technical nutrients. Biological nutrients will be designed to reenter ecological cycles to be consumed by microorganisms and other creatures in the soil. In addition to organic waste from our food, most packaging (which makes up about half the volume of our solid-waste stream) should be composed of biological nutrients. With today's technologies, it is quite feasible to produce packaging that can be tossed into the compost bin to biodegrade. /240

- In the United States, which is not a world leader in recycling, more than half of its steel is now produced from scrap. Similarly, there are more than a dozen paper mills running only on waste paper in the state of New Jersey alone. The new steel mini-mills do not need to be located near mines, nor the paper mills near forests. They are located near the cities that produce the waste and consume the raw materials, which saves considerable transportation costs. /241

- Many other ecodesign technologies for the repeated use of technical nutrients are on the horizon. For example, it is now possible to create special types of ink that can be removed from paper in a hot water bath without damaging the paper fibers. This chemical innovation would allow complete separation of paper and ink so that both can be reused. The paper would last ten to thirteen times longer than conventionally recycled paper fibers. If this technique were universally adopted, it could reduce the use of forest pulp up to 90 percent, in addition to reducing the amounts of toxic ink residues that now end up in landfills. /241

- If the concept of technical cycles were fully implemented, it would lead to a fundamental restructuring of economic relationships. After all, what we want from a technical product is not a sense of ownership but the service the product provides. We want entertainment from our VCR, mobility from our car, cold drinks from our refrigerator, and so on. As Paul Hawken likes to point out, we do not buy a television set in order to own a box of 4,000 toxic chemicals; we do so in order to watch television. /241

- From the perspective of ecodesign, it makes no sense to own these / products and to throw them away at the end of their useful lives. It makes much more sense to buy their services, i.e. to lease or rent them. Ownership would be retained by the manufacturer, and when one had finished using a product, or wanted to upgrade to a newer version, the manufacturer would take the old product back, break it down into its basic components—the technical nutrients—and use those in the assembly of new products, or sell them to other businesses. The resulting economy would no longer be based on the ownership of goods but would be an economy of service and flow. Industrial raw materials and technical components would continually cycle between manufacturers and users, as they would between different industries. /242

- This shift from a product-oriented economy to a service-and-flow economy is no longer pure theory. One of the world's largest carpet manufacturers, a company called Interface, based in Atlanta, has begun the transition from selling carpets to leasing carpenting services. The basic idea is that people want to walk on and look at a carpet, not own it. They can obtain those services at much lower cost if the company owns the carpet and remains responsible for keeping it in good shape in exchange for a monthly fee. Interface carpets are laid in the form of tiles, and only tiles that are worn are replaced after a regular monthly inspection. This reduces not only the amount of carpet material needed for replacements, but also minimizes disruptions, because the worn tiles are usually not found under furniture. When a customer wants to replace the entire carpet, the company takes it back, extracts its technical nutrients, and provides the customer with a new carpet in the desired color, style, and texture. /242

- Canon has revolutionized the photocopying industry by redesigning its copiers so that more than 90 percent of their components can be reused or recycled. /242

- In Fiat's Auto Recycling (FARE) system, the steel, plastics, glass, seat padding, and many other components of old FIAT cars are retrieved in over 300 dismantling centers, to be reused in new cars or passed on as resources to other industries. The company has established a target of 85 percent recycling of materials by 2002 and of 95 percent by 2010. The Fiat program has also been extended from Italy to other European countries and to Latin America. /243

- Another effect of this new product design will be to align the interests of manufacturers and customers when it comes to product durability. In an economy based on selling goods, the obsolence and frequent disposal and replacement of those goods is in the manufacturer's financial interests, even though that it is harmful to the environment and costly for the customers. In a service-and-flow economy, by contrast, it is in the interest of both manufacturers and customers to create long-living products while using a minimum of energy and materials. /243

- Natural Capitalism, by Paul Hawken, Amory Lovins, and Hunter Lovins, is full of astounding examples of dramatic increases in resource efficiency. The authors estimate that by pursuing these efficiencies we could almost halt the degradation of the biosphere, and emphasize that the present massive inefficiencies almost always cost more than the measures that would reverse them. /244

- A well-designed commercial structure will display a physical shape and orientation that takes the greatest advantage of the sun and wind, optimizing passive solar heating and cooling. That alone will usually save about one third of the building's energy use. Proper orientation, combined with other passive solar design features, also provides glare-free natural light throughout the structure, which usually provides sufficient lighting during daytime. /244

- Photovoltaic electricity can now be generated from wall panels, roofing shingles, and other structural elements that look and work like ordinary building materials but produce electricity whenever there is sunlight, even if it comes through clouds. A building with such photovoltaic materials as roofs and windows can produce more daytime electricity than it uses. Indeed, that is what half a million solar-powered homes around the world do every day. /245

- In a sustainable society, all human activities and industrial processes must ultimately be fueled by solar energy, like the processes in nature's ecosystems. Solar energy is the only kind of energy that is renewable and environmentally benign. Hence, the shift to a sustainable society centrally includes a shift from fossil fuels—the principal energy sources of the Industrial Age—to solar power. /247

- An estimated half a million homes around the world, mostly in remote villages that are not linked to an electric grid, now get their energy from solar cells. The recent invention of solar roofing tiles in Japan promises to lead to a further boost in the use of photovoltaic electricity. [T]hese 'solar shingles' are capable of turning rooftops into small power plants, which is likely to revolutionize electricity generation. /249

- Like many other products of industrial design, the contemporary automobile is stunningly inefficient. Only 20 percent of the energy in the fuel is used to turn the wheels, while 80 percent is lost in the engine's heat and exhaust. Moreover, a full 95 percent of the energy that is used moves the car, and only 5 percent moves the driver. The overall efficiency in terms of the proportion of fuel energy used to move the driver is 5 percent of 20 percent—a mere 1 percent! /252

- The differences between the physical properties of steel and fiber composites profoundly affect not only the design and operation of hypercars but also their manufacture, distribution, and maintenance. Although carbon fibers are more expensive than steel, the production process of composite car bodies is much more economical. Steel must be pounded, welded, and finished; composites emerge from a mold as a single, finished piece. This cuts tooling costs by up to 90 percent. The car assembly, too, is much simpler, since the lightweight parts are easy to handle and can be lifted without hoists. Painting, which is the most expensive and most polluting step in car manufacture, can be eliminated by integrating color into the molding process. /253-254

- The multiple advantages of fiber composites combine to favor small design teams, low break-even volumes per model, and local factories, all of which are characteristics of ecodesign as a whole. Maintenance of hypercars is also vastly simpler than that of steel cars, since many of the parts that are frequently responsible for mechanical breakdowns are no longer there. The rust-and-fatigue-free composite bodies, which are almost impossible to dent, will last for decades until they are eventually recycled. /254

- Another fundamental innovation is the hybrid-electric drive. Like other electric cars, hypercars have efficient electric motors to turn their wheels, as well as the ability to transform braking energy back into electricity, which offers additional energy savings. Unlike standard electric cars, however, hypercars have no batteries. Instead of using batteries, which continue to be heavy and short-lived, electricity is generated by a small engine, turbine or fuel cell. Such hybrid drive systems are small, and since they are not directly coupled to the wheels, they run near their optimal conditions all the time, which further reduces fuel consumption. /254

- The cleanest, most efficient and most elegant way to power a hybrid car is to use hydrogen in a fuel cell. Such an automobile not only operates silently and without any pollution, but also becomes, in effect, a small power plant on wheels. This is perhaps the most surprising and far-reaching aspect of the hypercar concept. When the car is parked at the owner's home or place of work—in other words, most of the time—the electricity produced by its fuel cell could be sent into the electric grid and the owner could automatically be credited for it. /254

- Perverse Subsidies include the billions of dollars paid by Germany to subsidize the extremely harmful coal-burning plants of the Ruhr Valley; the huge subsidies the U.S. government gives to its automobile industry, which was on corporate welfare during most of the twentieth century; the subsidies given to agriculture by the OECD, totaling $300 billion per year, which is paid to farmers to not grow food although millions in the world go hungry; as well as the millions of dollars the United States offers to tobacco farmers to grow a crop that causes disease and death. All of these are perverse subsidies indeed. They are powerful forms of corporate welfare that send distorted signals to the markets. Perverse subsidies are not officially tallied by any government in the world. While they support inequity and environmental degradation, the corresponding life-enhancing and sustainable enterprises are portrayed by the same governments as being uneconomical. It is high time to eliminate these immoral forms of government support. /258

- Another kind of signal the government sends to the marketplace is provided by the taxes it collects. At present, these too are highly distorted. Our existing tax systems place levies on the things we value—jobs, savings, investments— and do not tax the things we recognize as harmful—

pollution, environmental degradation, resource depletion, and so on. Like perverse subsidies, this provides investors in the marketplace with inaccurate information about costs. We need to reverse the system: instead of taxing incomes and payrolls, we should tax non-renewable resources, especially energy, and carbon emissions. /258

- To be successful, tax shifting needs to be a slow, long-term process in order to give new technologies and consumption patterns sufficient time to adapt, and it needs to be implemented predictably in order to encourage industrial innovation. Such a long-term, incremental shift of taxation will gradually drive wasteful, harmful technologies and consumption patterns out of the market. /259

- The analysis of living systems in terms of four interconnected perspectives—form, matter, process, and meaning—makes it possible to apply a unified understanding of life to phenomena in the realm of matter, as well as to phenomena in the realm of meaning. For example, we saw that metabolic networks in biological systems correspond to networks of communications in social systems; chemical processes producing material structures correspond to thought processes producing semantic structures; and flows of energy and matter correspond to flows of information and ideas. /261

- The so-called 'global market' is really a network of machines programmed according to the fundamental principle that money-making should take precedence over human rights, democracy, environmental protection, or any other value. /262

- The United States projects its tremendous power around the world to maintain optimal conditions for the perpetuation and expansion of production. The central goal of its vast empire—its overwhelming military might, impressive range of intelligence agencies, and dominant positions in science, technology, media, and entertainment—is not to expand its territory, nor to promote freedom and democracy, but to make sure that it has global access to natural resources and that markets around the world remain open to its products. /263

- This glorification of material consumption has deep ideological roots that go far beyond economics and politics. Its origins seem to lie in the universal association of manhood with material possessions in patriarchal cultures. /264

Chapter Ten

Steering Business Toward Sustainability

Fritjof Capra

Steering Business Toward Sustainability

Edited with Wolfgang Pauli
New York: United Nations University Press, 1995

Review

Steering Business Toward Sustainability is a book of high practical value for leaders and organizations who are conscious of the need for deep ecology and the challenge we presently face to update most of our basic business routines and procedures in order to build sustainable organizations.

> Quite simply, our business practices are destroying life on earth. Given current corporate practices, not one wildlife reserve, wilderness, or indigenous culture will survive the global market economy. /1

Capra's idea of ecology has developed over many years. It is rooted in the insights he exposed in his previous four books, and thus we can say this present book is solidly grounded in research. In addition, Capra leaves no doubt that it's not just a technocratic idea, but an intrinsically spiritual concept. He also credits those, religions and peoples, who have practiced ecological thinking long before the birth of the United States of America:

> When the concept of the human spirit is understood as the mode of consciousness in which the individual feels connected to the cosmos as a whole, it becomes clear that

ecological awareness is spiritual in its deepest essence. It is therefore not surprising that the emerging new vision of reality, based on deep ecological awareness, is consistent with the so-called perennial philosophy of spiritual traditions, whether we talk about the spirituality of Christian mystics, that of Buddhists, or the philosophy and cosmology underlying the American Indian traditions. /3

Capra reminds us of the fact that when restructuring our economies, we should learn from nature, instead of feeling superior over nature. *Ecoliteracy* is one of the notions Capra is currently lecturing about, and Gunter Pauli, the co-editor of this reader is one of Capra's truest collaborators, himself an authority on ecology in Germany. Within the concept of ecological literacy, Capra seems to give the highest importance to the term *sustainability*, and he comprehensively explains what this term means:

> In our attempts to build and nurture sustainable communities we can learn valuable lessons from ecosystems, because ecosystems are sustainable communities of plants, animals, and microorganisms. To understand these lessons, we need to learn nature's language. We need to become ecologically literate. (...) Being ecologically literate means understanding how ecosystems organize themselves so as to maximize sustainability. /4

Many of us have yet to understand why our modern technologies are so much in conflict with nature's setup, and this is a fact that is barely elucidated in the mass media. Non-educated people, and even entrepreneurs who have not been exposed to academic study are usually at pains with understanding the deeper reasons of this conflict. Capra, referencing Paul Hawken, *The Ecology of Commerce*, Harper, 1993, elucidates it:

> The present clash between business and nature, between economics and ecology, is mainly due to the fact that nature is cyclical, whereas our industrial systems are linear, taking up energy and resources from the earth, transforming them into products plus waste, discarding the waste, and finally throwing away also the products after they have been used. Sustainable patterns of production and consumption need to be cyclical, imitating the processes in ecosystems./5

Back in Antiquity, there was barely a need for people to learn systems thinking because they were naturally aligned with the logic of nature, as they were living *with* nature, and not on top of nature, as we do today. We can also say that we as modern city dwellers have lost our continuum, as it was expressed with much emphasis by Jean Liedloff in *The Continuum Concept*. Besides, Capra informs us about how we should apply ecology in our daily lives, and what it teaches us. There are seven principles to learn that Capra calls *Principles of Ecology* and that he explains one by one:

> *Interdependence*
> All members of an ecosystem are interconnected in a web of relationships, in which all life processes depend on one another.
>
> *Ecological Cycles*
> The interdependencies among the members of an ecosystem

involve the exchange of energy and resources in continual cycles.

Energy Flow
Solar Energy, transformed into chemical energy by the photosynthesis of green plants, drives all ecological cycles.

Partnership
All living members of an ecosystem are engaged in a subtle interplay of competition and cooperation, involving countless forms of partnership.

Flexibility
Ecological cycles have the tendency to maintain themselves in a flexible state, characterized by interdependent fluctuations of their variables.

Diversity
The stability of an ecosystem depends on the degree of complexity of its network of relationships; in other words, on the diversity of the ecosystem.

Coevolution
Most species in an ecosystem coevolve through an interplay of creation and mutual adaptation.

Sustainability
The long-term survival of each species in an ecosystem depends on a limited resource base. Ecosystems organize themselves according to the principles summarized above so as to maintain sustainability./6

Capra also explains very well the feedback-looping that we find is a typical feature of living systems. The understanding of feedbacking by *constant parameter change* as a response to a given stimulus is crucial for the understanding of the cyclic nature of all life. This is one of the points modern scientists are really at pains with because their thought structure simply is too linear. Capra explains:

> When changing environmental conditions disturb one link in an ecological cycle, the entire cycle acts as a self-regulating feedback loop and soon brings the situation back into balance. And since these disturbances happen all the time, the variables in an ecological cycle fluctuate continually. These fluctuations represent the ecosystem's flexibility. Lack of flexibility manifests itself as stress. In particular, stress will occur when one or more variables of the system are pushed to their extreme values, which induces increased rigidity throughout the system. Temporary stress is an essential aspect of life, but prolonged stress is harmful and destructive to the system./7

It's exactly this widely unpredictable feedback-looping that is inherent in the current paradigm of ecological destruction. This dangerous situation is worsened by the general lack of ecological literacy regarding the possible effects of large disturbances, such as ozone hole, deforestation, global warming and desertification.

Our knowledge also is insufficient to make ecological solutions work *effectively* even once ecology-friendly policies are implemented by governments and organizations. It is not enough to see the dangers and implement good new laws for protecting nature, we also need to see how the damage already done will interact with our new policies; this is so because it's not taken for granted that our best-intended tactics of healing nature are going to work. For insuring this, we have to learn much more about feedback-looping in natural systems, and we need to learn how nature heals herself.

For example, it has been shown that the planting of new trees does *not* per see heal the damage that deforestation has done to our planet. It's all in the why and how of planting trees, where, how many, and in what mixture of species that the wisdom lies. On the other hand, it has been seen in Indonesia, one of the worst hit countries by deforestation, that huge areas that were deforested began to grow trees without anybody doing anything about it! Later research showed that the conditions had been ideal for trees to grow again, but nobody really knew why in other places, where at first sight conditions were very similar, this was not the case.

We definitely have to develop humility, given our dreadful ignorance in the face of the complexity level of nature, at all phases of evolution. We are simply not trained in complexity thinking, and our schools and universities destroy the little of complexity we have developed naturally as children as a result of free play. It is *freedom* that is at the basis of building complexity, not discipline, it is permissiveness, not repression. Here is where our morality clearly stares grimly in nature's face because nature is amoral. If theologians will ever grasp this dimension is not my concern, but as scientists we should definitely do away with our projections upon nature and at the same time get all our senses and our emotional intelligence ready for receiving the messages of nature. Nature communicates when we are ready to listen, and it will tell us how we can help healing the damage we have done to her over the last five thousand years of patriarchal ignorance.

This book together with *Hidden Connections* and *The Web of Life* teaches the basics of understanding nature's complexity. It also teaches us the importance of *diversity*, a concept that at present is rather shunned by mainstream politics, while liberal phases, as it was the case through the 1970s, foster higher levels of cultural diversity. Nature shows us that this is not just a random development but that it's diversity on which side is intelligent and life-fostering behavior, and not uniformity. This is so, inter alia, because diversity fosters flexibility, and vice versa, while uniformity entails rigidity. What does loss of biodiversity on the planet mean for our future as a human race? The regard here is rather dim, and Capra leaves no doubt about it:

> In ecosystems, flexibility through fluctuations does not always work, because there can be very severe disturbances that actually wipe out an entire species. In other words, one of the links in the ecosystem's network is destroyed. An ecological community will be resilient when this link is not the only one of its kind; when there are other connections that can at least partially fulfill its functions. In other words, the more complex the network, the greater the diversity of

its interconnections, the more resilient it will be. The same is true in human communities. Diversity means many different relationships, many different approaches to the same problem. A diverse community is a resilient community, capable of adapting easily to changing situations./8

The loss of biodiversity, i.e. the daily loss of species, is in the long run one of our most severe global environmental problems. And because of the close integration of tribal indigenous people into their ecosystems, the loss of biodiversity is closely tied to the loss of cultural diversity, the extinction of traditional tribal cultures. This is especially important today. As the beliefs and practices of the industrial culture are being recognized as part of the global ecological crisis, there is an urgent need for a wider understanding of cultural patterns that are sustainable. The vast folk wisdom of American Indian, African, and Asian traditions has been viewed as inferior and backward by the industrial culture. It is time to reverse this Euro-centric arrogance and to recognize that many of these traditions—their ways of knowing, technologies, knowledge of foods and medicines, forms of aesthetic expression, patterns of social interaction, communal relationships, etc.—embody the ecological wisdom we so urgently need today./8

This is what I am saying since about twenty years, having founded, back in 1994, *Ayuda International Foundation* for the protection of tribal people's wisdom about life, and their high cultural diversity, and wistful traditions for living in alignment with the laws of nature.

Yet it's a fact that in most developing countries technologies for recycling and for healing the badly afflicted metropoles are costly and not as accessible and readily available as in wealthy high-tech nations. Only truly supportive cultural and technological exchange between rich and poor countries can help changing this dim picture. Whatever our personal opinions are in the face of these huge global problems, that also our next generations will be burdened with, we have to keep an open mind and learn, and change our rigid positions.

Fritjof Capra and Wolfgang Pauli have given in this reader very useful suggestions that can be taken as starting points for deeper study, as the field of investigation is huge, and never-ending. Nature's complexity is perhaps the single most important topic of study for 21st century science, and I hope I can contribute a little to it by my own efforts. As for the authors of this book, they surely have done their very substantial contributions!

Quotes

- Quite simply, our business practices are destroying life on earth. Given current corporate practices, not one wildlife reserve, wilderness, or indigenous culture will survive the global market economy. /3 (Referencing Paul Hawken, The Ecology of Commerce, New York: Harper Collins, 1993)

- The paradigm that is now receding has dominated Western industrial culture for several hundred years, during which it has shaped modern society and has significantly influenced all parts of the world. This paradigm consists of a number of ideas and values, among them the view of the universe as a mechanical system composed of elementary building blocks, the view of the human body as a machine, the view of life in society as a competitive struggle for existence, the belief in unlimited material progress to be achieved through economic and technological growth, and—last, not least—

the belief that a society in which the female is everywhere subsumed under the male is one that follows a basic law of nature. All these assumptions have been fatefully challenged by recent events. And, indeed, a radical revision of them is now occurring. /2-3

- When the concept of the human spirit is understood as the mode of consciousness in which the individual feels connected to the cosmos as a whole, it becomes clear that ecological awareness is spiritual in its deepest essence. It is therefore not surprising that the emerging new vision of reality, based on deep ecological awareness, is consistent with the so-called 'perennial philosophy' of spiritual traditions, whether we talk about the spirituality of Christian mystics, that of Buddhists, or the philosophy and cosmology underlying the American Indian traditions. /3

- Living systems include individual organisms, parts of organisms, and communities of organisms, such as social systems and ecosystems. All these are irreducible wholes whose specific structures arise from the interactions and interdependence of their parts. Systems theory tells us that all these living systems share a set of common properties and principles of organization. /4

- In our attempts to build and nurture sustainable communities we can learn valuable lessons from ecosystems, because ecosystems are sustainable communities of plants, animals, and microorganisms. To understand these lessons, we need to learn nature's language. We need to become ecologically literate. (...) Being ecologically literate means understanding how ecosystems organize themselves so as to maximize sustainability. /4

- The present clash between business and nature, between economics and ecology, is mainly due to the fact that nature is cyclical, whereas our industrial systems are linear, taking up energy and resources from the earth, transforming them into products plus waste, discarding the waste, and finally throwing away also the products after they have been used. Sustainable patterns of production and consumption need to be cyclical, imitating the processes in ecosystems. /5 (Quoting Hawken, pp. 62 ff.)

- In the nineteenth century, the Social Darwinists saw only competition in nature. Today we know that all competition takes place within a broader context of cooperation involving countless forms of partnership. Indeed, partnership—the tendency to associate, establish links, live inside one another and cooperate—is an essential characteristic of living organisms. /6

- [Principles of Ecology]
  Interdependence
  All members of an ecosystem are interconnected in a web of relationships, in which all life processes depend on one another.

- **Ecological Cycles**
  The interdependencies among the members of an ecosystem involve the exchange of energy and resources in continual cycles.

- **Energy Flow**
  Solar energy, transformed into chemical energy by the photosynthesis of green plants, drives all ecological cycles.

- **Partnership**
  All living members of an ecosystem are engaged in a subtle interplay of competition and cooperation, involving countless forms of partnership.

- **Flexibility**
  Ecological cycles have the tendency to maintain themselves in a flexible state, characterized by interdependent fluctuations of their variables.

- **Diversity**
  The stability of an ecosystem depends on the degree of complexity of its network of relationships; in other words, on the diversity of the ecosystem.

- **Coevolution**
  Most species in an ecosystem coevolve through an interplay of creation and mutual adaptation.

- **Sustainability**
  The long-term survival of each species in an ecosystem

depends on a limited resource base. Ecosystems organize themselves according to the principles summarized above so as to maintain sustainability. /6

- The general shift from domination to partnership is an essential part of the shift from the mechanistic to the ecological paradigm. Whereas a machine is properly understood through domination and control, the understanding of a living system will be much more successful if approached through cooperation and partnership. Cooperative relationships are an essential characteristic of life. /7

- The more variables are kept fluctuating, the more dynamic is the system, the greater is its flexibility, the greater its ability to adapt to changing environmental conditions. /7

- When changing environmental conditions disturb one link in an ecological cycle, the entire cycle acts as a self-regulating feedback loop and soon brings the situation back into balance. And since these disturbances happen all the time, the variables in an ecological cycle fluctuate continually. These fluctuations represent the ecosystem's flexibility. Lack of flexibility manifests itself as stress. In particular, stress will occur when one or more variables of the system are pushed to their extreme values, which induces increased rigidity throughout the system. Temporary stress is an essential aspect of life, but prolonged stress is harmful and destructive to the system. /7

- In ecosystems, flexibility through fluctuations does not always work, because there can be very severe disturbances that actually wipe out an entire species. In other words, one of the links in the ecosystem's network is destroyed. An ecological community will be resilient when this link is not the only one of its kind; when there are other connections that can at least partially fulfill its functions. In other words, the more complex the network, the greater the diversity of its interconnections, the more resilient it will be. The same is true in human communities. Diversity means many different relationships, many different approaches to the same problem. A diverse community is a resilient community, capable of adapting easily to changing situations. /8

- The loss of biodiversity, i.e. the daily loss of species, is in the long run one of our most severe global environmental problems. And because of the close integration of tribal indigenous people into their ecosystems, the loss of biodiversity is closely tied to the loss of cultural diversity, the extinction of traditional tribal cultures. This is especially important today. As the beliefs and practices of the industrial culture are being recognized as part of the global ecological crisis, there is an urgent need for a wider understanding of cultural patterns that are sustainable. The vast folk wisdom of American Indian, African, and Asian traditions has been viewed as inferior and backward by the industrial culture. It is time to reverse this Euro-centric arrogance and to recognize that many of these traditions—their ways of knowing, technologies, knowledge of foods and medicines, forms of aesthetic expression, patterns of social interaction, communal relationships, etc.—embody the ecological wisdom we so urgently need today. /8

- We believe that if the corporate world does not play an active role in redefining its own operations, moving toward sustainability, the world as a whole will never succeed in that task. /9

- Executives know too that doing more business the same way as in the past is a guarantee of failure. By the same token, the time is past when scholars, governments, or environmental organizations could hand down a doctrine from some high pulpit of academic certainty or from emotional distress and fears. Rather, cooperation among all those who have a stake in the future of society is critical. /9

- The company of the 21st century will have different responsibilities to assume in society from those generally expected today. (…) This implies that if business does not take a proactive and really credible leap towards converting its operations on the basis of the concept of ecological sustainability, it will lose its legitimacy. /12

Chapter Eleven

The Science of Leonardo

The Science of Leonardo

Inside the Mind of the Great Genius of the Renaissance
New York: Anchor Books, 2008
First published with Doubleday, 2007

Review

Fritjof Capra notes in his elucidating study on Leonardo, *The Science of Leonardo (2007/2008)*, that the great polymath of the Renaissance was contrary to common belief not a mechanistic thinker, as were later, for example, Francis Bacon or Galileo Galilei, despite the fact that he was one of the first great inventors of modern machines, and actually very interested in machines all his life through. But he did not, as later Cartesian science and philosophers such as La Mettrie or Baron d'Holbach, consider the human body as a machine.

The world was used to see *Leonardo da Vinci (1452-1519)* as a painter, not a scientist. I questioned this view already at the start of my genius research, about thirty years ago, when I found out about Leonardo's scientific notebooks. Leonardo and Goethe were the avatars of a new culture, a new society, and yet, at their lifetimes, their breadth of mind and holistic worldview was hardly valued, let alone understood. Goethe had a stable income as a government-employed jurist, Leonardo was doing work for kings and queens, and made a living with construing weapons, but both had their minds focused on what essentially constitutes life, and Leonardo, just as later Albert Einstein, was a *scientist* genius before he was a great artist. Before the 20th century, both scientists were barely understood. Goethe's color theory was looked at with suspicion, as it was in flagrant contradiction to Newton's scientific universe.

—Johann Wolfgang von Goethe, The Theory of Colors, New York: MIT Press, 1970, first published in 1810, Frederick Burwick, The Damnation of Newton: Goethe's Color Theory and Romantic Perception, New York, Walter de Gruyter, 1986 and Dennis L. Sepper, Goethe Contra Newton: Polemics and the Project of a New Science of Color: Cambridge: Cambridge University Press, 1988

Leonardo was considered by Herman Grimm, a noted historian, in side remarks of his monograph *Life of Michelangelo*, as a flamboyant regal person, but also a bohemian and 'dark soul.' Relating Vasari's autobiography he writes:

> Lionardo is not a man that you can pass at ease, but a force that we are bound with and whose charm we cannot escape when it once has touched us. Whoever has seen Mona Lisa smile, is followed eternally by this smile, just as by Lear's fury, Macbeth's ambition, Hamlet's depression or Iphigenia's moving purity.
>
> —Herman Grimm, Leben Michelangelos, Wien, Leipzig: Phaidon Verlag, 1901, 42 (Translation mine)

> It is as if Lionardo had within himself the need of the most daring contradictions in relation to the truly wonderful beings he was able to create. He himself, handsome, and strong as a titan, generous, surrounded with numerous servants and horses, and fantastic household, a perfect musician, charming and lovely in sight of high and low, poet, sculptor, architect, civil engineer, mechanic, a friend of counts and kings and yet as citizen of his nation a dark existence who, seldom leaving the semi-dark atmosphere of his being, finds no opportunity to invest his forces simply and freely for a great endeavor. (Id., 43-44)

Such natures, that with their extraordinary talents seem to be born only for adventure and who have kept even in the most serious and deepest endeavors of their mind a child-like playfulness, are rare, but possible appearances. Such men are of high descent; beautiful, independent and glowing of yet undefined action, they walk into the world. All is open to them and in no way they encounter real, oppressive sorrow; they mold their lives that nobody than themselves understands because nobody has been born under conditions that exactly led to such a fantastic yet necessary and inescapable destiny. (Id., 44)

Grimm's picture of Leonardo lacks personal touch; it is deeply romantic and seems almost sterile. Grimm did not depict, and even less appreciate, the personal identity of the *genius* but rather painted him as a *genus*. Needless to add that in his romantic effluvia, Grimm did not lose a word on the *scientist* Leonardo, and this is all too typical for the general opinion about him before the 20th century. Now, with the study of his scientific genius by Fritjof Capra, Leonardo can eventually be noted by science history as one of the greatest scientific innovators the world has ever seen.

Capra makes his point convincingly that modern science did not begin with Galilei, but with Leonardo, because it was Leonardo who, for the first time in human history, has applied the *scientific method,* logic, observation and the capacity to conceptualize a multitude of single data into a single coherent and consistent theory. This was so much the more an achievement as during his lifetime science was still entangled with religion to a point that a large body of the *corpus scientia* was ecclesiastical doctrine, and as such a mix of mythic views, politically correct assumptions and a residue of observation that was for the largest part taken over from Aristotle. Capra writes:

> Leonardo da Vinci broke with this tradition. One hundred years before Galileo and Bacon, he single-handedly developed a new empirical approach to science, involving the systematic observation of nature, logical reasoning, and some mathematical formulations—the main characteristics of what is known today as the *scientific method.* /2

It is highly curious to observe that Leonardo did not formulate, at the onset of his lifelong multidisciplinary research, an intention for so doing; calling himself humbly 'uomo senza lettere,' an uneducated man, his project was to write a manual on the 'science of painting.' His grasp of the world was predominantly visual, and so was his scientific method; it was primarily based upon very accurate and very astute observation of nature and all forms of living. Only a genius can have the abundant curiosity, the intellectual grasp and the persistence to inquire so deeply and so thoroughly from what the eye perceives, to really get to unveil basic laws and functional connections in all living, and in all material life.

One may be baffled to see that this magnificent creator was to that point marginalized during his lifetime that none of his notebooks were ever published, worse, as Capra reports, after his death, the collection of his writings and drawings, almost thirteen thousand pages, was scattered and dispersed all over Europe, and stuffed in libraries, instead of having been sorted and properly published; still worse, almost half of the collection was lost. Capra writes:

> Leonardo's scientific work was virtually unknown during his lifetime and remained hidden for over two centuries after his death in 1519. His pioneering discoveries and ideas had no direct influence on the scientists who came after him, although during the subsequent 450 years his conception of a science of living forms would emerge again at various times. (…) While Leonardo's manuscripts were gathering dust in ancient European libraries, Galileo Galilei was being celebrated as the 'father of modern science.' I cannot help but argue that the true founder of modern science was Leonardo da Vinci, and I wonder how Western scientific thought would have developed had his Notebooks been known and widely studied soon after his death. /5-6

I would like to focus for a moment on one single and in my view *significant* detail, namely how Leonardo was thinking about 'life,' about living systems, and about science in relation to life. We are today familiar with the conception of life being not a linear rigid structure that is totally measurable, except when organisms have died, but a nonlinear structure of dynamic patterns, which are essentially *relationships*. As we have seen, Fritjof Capra has elucidated in his study *The Web of Life (1997)* that life is basically a structure of 'networks within networks' and that hierarchies do exist in nature only in the sense that smaller networks are contained in larger networks but not in the sense of a rigid up-down hierarchy as traditional human society, especially under patriarchy, has conceptualized it as the reigning sociopolitical model.

This view is emerging since a few decades and is called the 'systems view of life;' it is related to *deep ecology* and *Gaia theory* and was developed, besides Capra, mainly by Ludwig von Bertalanffy, Humberto Maturana, Francisco J. Varela, Ilya Prigogine and Ervin Laszlo.

What was known from Goethe's pantheistic philosophy that considered life as an organic whole, we find it, in Capra's retrospective, equally with Leonardo. Capra writes:

> Nature as a whole was alive for Leonardo. He saw the patterns and processes in the microcosm as being similar to those in the macrocosm. (…) / While the analogy between microcosm and macrocosm goes back to Plato and was well known throughout the Middles Ages and the Renaissance, Leonardo disentangled it from its original mythical context and treated it strictly as a scientific theory. / 3-4

Capra goes as far as talking of Leonardo as 'a systemic thinker,' because of his strong synthetic thinking ability, that was able to 'interconnect observations and ideas from different disciplines.' / 5

He observes that Leonardo's visual perception was unusually sharp and accurate, and truly scientific in scope and intent, and that he also had an accurate sense of *motion* which is seldom to find. Usually, the static eye distorts objects that are in motion. We are hardly aware of this imperfection of our sight as we today are surrounded by visual objects such as televisions, and take high-quality photographs using digital technology. But at a time when there were *no photographic plates and cameras,* motion was hardly ever depicted by visual artists in a realistic sense; this was simply so as most artists were unable to train their eye to a point to perceive motion correctly, and without distortion of perspective.

In addition, Capra notes, Leonardo had a view of the body that preceded quantum physics and modern spirituality. For Leonardo, 'the human body was an outward and visible expression of the soul; it was shaped by its spirit.' /5

> Unlike Descartes, Leonardo never thought of the body as a machine, even though he was a brilliant engineer who designed countless machines and mechanical devices. /Id.

Fritjof Capra notes that Leonardo had an understanding of nature that was basically *ecological* in the sense that, contrary to what Francis Bacon would advocate a century later, man was not made for dominating nature, but for *understanding* nature, and based upon that understanding, to *cooperate* with nature. From this basic worldview, Leonardo was sensible to nature's complexity and abundance, which was certainly not an attitude commonly to be found at his lifetime. In addition, he was aware of the fallacy of scientific reductionism. Capra notes:

Our sciences and technologies have become increasingly narrow in their focus, and we are unable to understand our multifaceted problems from an interdisciplinary perspective. /12

We urgently need a science that honors and respects the unity of all life, that recognizes the fundamental interdependence of all natural phenomena, and reconnects us with the living earth. What we need today is exactly the kind of thinking and science Leonardo da Vinci anticipated and outlined five hundred years ago, at the height of the Renaissance and the dawn of modern scientific age. /Id.

Now, about the specific notion of genius, that we must take for granted, for the *nature of genius* has been discussed both in philosophy and modern science for quite a time.

—See Peter Fritz Walter, Creative Genius: Four-Quadrant Creativity in the Lives and Works of Leonardo da Vinci, Wilhelm Reich, Albert Einstein, Svjatoslav Richter and Keith Jarrett (2014).

Fritjof Capra elucidates comprehensively the fact that our modern notion of genius, in the scientific sense as defined by creativity research and neuroscience has little to do with what the ancients believed was the origin of genius:

> The Latin word *genius* originated in Roman religion, where it denoted the spirit of the *gens*, the family. It was understood as a guardian spirit, first associated with individuals and then also with peoples and places. The extraordinary achievements of artists or scientists were attributed to their genius, or attendant spirit. This meaning of genius was prevalent throughout the Middle Ages and the Renaissance.

In the eighteenth century, the meaning of the word changed to its familiar modern meaning to denote these individuals themselves, as in the phrase 'Newton was a genius.' /28

I will end my review here for this book is so particular and detailed that I would need to paraphrase too much of Capra's good and competent narration. This book and his next book, *Learning from Leonardo (2013)*, are very great achievements of the writer and scientific thinker Fritjof Capra. His excellent mastery of the Italian language understandably was extremely helpful to him in perusing—or rather *deciphering*—Leonardo's shorthand writing style.

Quotes

- Today, from the vantage point of twenty-first-century science, we can recognize Leonardo da Vinci as an early precursor of an entire lineage of scientists and philosophers whose central focus was the nature of organic form. /5

- In the Italian Renaissance, the association of exceptional creative powers with divine inspiration was expressed in a very direct way by bestowing on those individuals the epithet *divino*. Among the Renaissance masters, Leonardo as well as his younger contemporaries Raphael and Michelangelo were acclaimed as divine. /29

- Closely associated with the powers of intense concentration that are characteristic of geniuses seems to be their ability to memorize large amounts of information in the form of a coherent whole, a single gestalt. /30

- The intellectual climate of the Renaissance was decisively shaped by the philosophical and literary movement of humanism, which made the capabilities of the human individual its central concern. This was a fundamental shift from the medieval dogma of understanding human nature from a religious point of view. /32

- For Leonardo ... being universal meant to recognize similarities in living forms that interconnect different facets of nature—in this case, anatomical structures of different animals. The recognition that nature's living forms exhibit such fundamental patterns was a key insight of the school of Romantic biology in the eighteenth century. These patterns were called *Urtypen* ('archetypes') in Germany, and in England Charles Darwin acknowledged that his concept played a central role in his early conception of evolution. In the twentieth century, anthropologist and cyberneticist Gregory Bateson expressed the same idea in the succinct phrase 'the pattern which connects.' Thus, Leonardo da Vinci was the first in a lineage of scientists who focused on the patterns interconnecting the basic structures and processes of living systems. Today, this approach to science is called 'systemic thinking.' /34

- Leonardo's anatomical drawings were so radical in their conception that they remained unrivaled until the end of the eighteenth century, nearly three hundred years later. Indeed, they have been praised as the beginning of modern anatomical illustration. /39

- According to complexity theory, creativity—the generation of new forms—is a key property of all life, and it involves the very process that Leonardo revealed in his exquisite preparatory drawings. I would argue that our most creative insights emerge from such states of uncertainty and confusion. /42

- At the beginning of the Renaissance, painting was classified as a 'mechanical art,' together with crafts like gold and metal work, jewelry, tapestry, and embroidery. None of these mechanical arts stood out in terms of prestige, and their practitioners remained relatively anonymous. /43

- What made Leonardo unique as a designer and engineer, however, was that many of the novel designs he presented in his Notebooks involved technological advances that would not be realized until several centuries later. /54

- Italy in the fifteenth century was a kaleidoscope of over a dozen independent states, which formed ever-shifting alliances in a constant struggle for economic and political power that was always on the verge of degenerating into war. The principle powers of the time were the duchies of Milan and Savoy and the republic of Venice in the north, the republic of Florence and the territories of the papacy in the center of the peninsula, and the kingdoms of Naples and Sicily in the south. In addition, there were a number of smaller states—Genoa, Mantua, Ferrara, and Siena. /65

- Leonardo had to move many times in the face of impending war, foreign occupations, and other changes of political power. Thus the trajectory of his life led him from Florence to Milan, from Milan to Venice, back to Florence, back again to Milan, then to Rome, and finally to Amboise in France. /Id.

- In Vinci, Leonardo attended one of the customary *scuole d'abaco* ('abacus schools'), which taught children reading, writing, and a rudimentary knowledge of arithmetic adapted to the needs of merchants. Students who prepared for university then moved on to a *scuola di lettere* ('schools of letters'), where they were taught the humanities based on the study of the great Latin authors. Such an education included rhetoric, poetry, history, and moral philosophy. /67

- Being an illegitimate child, Leonardo was barred from attending university, and hence was not sent to a scuola di lettere. Instead he began his apprenticeship in the arts. /67

- The six years Leonardo spent at the French court in Milan marked a stage of maturity both in his science and his art. During those years the artist slowly developed and refined three of his mature master paintings: the *Madonna and Child with Saint Anne*, the *Leda*, and his most famous painting, the *Mona Lisa*. /113

- The Château d'Amboise was the home of French kings and queens for over 150 years. François I had spent his childhood and youth there, and used it as his principal residence. /126

- Just as Alexander the Great, another young warrior-king, had been tutored by Aristotle, the great philosopher of antiquity, so François I was now tutored by Leonardo da Vinci, the great sage and genius of the Renaissance. /126-127

- During his time at Amboise, Leonardo also advised the king on various architectural and engineering projects, in which he revived his conception of buildings and cities as 'open systems' (to use our modern term), in which people, material goods, food, water, and waste need to move and flow easily for the system to remain healthy. /129

- After Melzi's death in 1570, his son Orazio, who did not share his father's reverence for the great Leonardo, carelessly stuffed the Notebooks into several chests in the villa's attic. When it became known that batches of Leonardo's exquisite drawings could easily be obtained from Orazio, souvenir hunters turned up at Vaprio; they were allowed to take whatever they wanted. Pompeo Leoni of Arezzo, sculptor at the court of Madrid, obtained close to fifty bound volumes in addition to about two thousand single sheets, which he took to Spain in 1590. Thus, at the turn of the sixteenth to the seventeenth century, Spain had the largest concentration of Leonardo's writings and drawings. /131

- In contrast to the Romans, the Arab scholars not only assimilated Greek knowledge but examined it critically and added their own commentaries and innovations. Numerous editions of these texts were housed in huge libraries throughout the Islamic empire. In Moorish Spain, the great library of Córdoba alone contained some six hundred thousand manuscripts. /139

- Islamic religious leaders emphasized compassion, social justice, and a fair distribution of wealth. Theological speculations were seen as being far less important and therefore discouraged. As a result, Arab scholars were free to develop philosophical and scientific theories without fear of being censored by their religious authorities. /Id.

- The dark side of this seamless fusion of science and theology was that any contradiction by future scientists would necessarily have to be seen as heresy. In this way, Thomas Aquinas enshrined in his writings the potential for conflicts between science and religion—which indeed arose three centuries later in Leonardo's anatomical research, reached a dramatic climax with the trial of Galileo, and have continued to the present day. /140

- By 1600 the surface of the known world had doubled since medieval times. /143

- Florence under the Medici was the center of Platonism. Milan, under the influence of the universities of Padua and Bologna, was predominantly Aristotelian. /148

- At the core of Hippocratic medicine was the conviction that illnesses are not caused by supernatural forces, but are natural phenomena that can be studied scientifically and influenced by therapeutic procedures and wise management of one's life. Thus medicine should be practiced as a scientific discipline and should include the prevention of illness, as well as its diagnosis and treatment. This attitude has formed the basis of scientific medicine to the present day. /153

- Twentieth-century science has shown repeatedly that all natural phenomena are ultimately interconnected, and that their essential properties, in fact, derive from their relationships to other things. Hence, in order to explain any one of them completely, we would have to understand all the others, which is obviously impossible. This insight has forced us to abandon the Cartesian belief in the certainty of scientific knowledge and to realize that science can never provide complete and definite explanations. In science, to put it bluntly, we never deal with truth, in the sense of a precise correspondence between our descriptions and the described phenomena. We always deal with limited and approximate knowledge. /159

- Five hundred years before the scientific method was recognized and formally described by philosophers and scientists, Leonardo da Vinci single-handedly developed and practiced its essential characteristics—study of the available literature, systematic observations, experimentation, careful and repeated measurements, the formulation of theoretical models, and frequent attempts at mathematical generalizations. /Id.

- What turned Leonardo from a painter with exceptional gifts of observation into a scientist was his recognition that his observations, in order to be scientific, needed to be carried out in an organized, methodical fashion. Scientific experiments are performed repeatedly and in varying circumstances so as to eliminate accidental factors and technical flaws as much as possible. The parameters of the experimental setting are varied in order to bring to light the essential unchanging features of the phenomena being investigated. This is exactly what Leonardo did. He never tired of carrying out his experiments and observations again and again, with fierce attention to the minutest details, and he would often vary his parameters systematically to test the consistency of his results. /162

- Such detailed studies of vortices in turbulent water were not taken up again for another 350 years, until the physicist Hermann von Helmholtz developed a mathematical analysis of vortex motion in the mid-nineteenth century. /175

- Whenever Leonardo explored the forms of nature in the macrocosm, he also looked for similarities of patterns and processes in the human body. In so doing, he went beyond the general analogies between macro- and microcosm that were common knowledge at his time, drawing parallels between very sophisticated observations in both realms. He applied his knowledge of turbulent flows of water to the movement of blood in the hearts and aorta. He saw the 'vital sap' of plants as their essential life fluid and observed that it nourishes the plant tissues, as the blood nourishes the tissues of the human body. He noticed the structural similarity between the stalk (known to botanists as the funiculus) that attaches the seed of the plant to the tissues of the fruit, and the umbilical cord that attaches the human fetus to the placenta. /177

- Leonardo's observation was restated by Isaac Newton two hundred years later and has since been known as Newton's third law of motion. /185

- When we look at Leonardo's geometry from the point of view of present-day mathematics, and in particular from the perspective of complexity theory, we can see that he developed the beginnings of the branch of mathematics known as topology. /206-207

- Like Galileo, Newton, and subsequent generations of scientists, Leonardo worked from the basic premise that the physical universe is fundamentally ordered and that its causal relationships can be comprehended by the rational mind and expressed mathematically. /210

- Since Leonardo's science was a science of qualities, of organic forms and their movements and transformations, the mathematical 'necessity' he saw in nature was not one expressed in quantities and numerical relationships, but one of geometric shapes continually transforming themselves according to rigorous laws and principles. /Id.

- This meaning of 'mathematical' is quite different from the one understood by scientists during the Scientific Revolution and the subsequent three hundred years. However, it is not unlike the understanding of some of the leading mathematicians today. The recent development of complexity theory has generated a new mathematical language in which the dynamics of complex systems—including the turbulent flows and growth patterns of plants studied by Leonardo—are no longer represented by algebraic relationships, but instead by geometric shapes, like the computer-generated strange attractors or fractals, which are analyzed in terms of topological concepts. /210-211

- Like Leonardo da Vinci five hundred years ago, modern mathematicians today are showing us that the understanding of patterns, relationships, and transformations is crucial to understand the living world around us, and that all questions of pattern, order, and coherence are ultimately mathematical. /211

- To a modern intellectual, used to the exasperating fragmentations of academic disciplines, it is amazing to see how Leonardo moved swiftly from perspective and the effects of light and shade to the nature of light, the pathways of optic nerves, and the actions of the soul. /213

Chapter Twelve

Learning from Leonardo

Learning from Leonardo

Decoding the Notebooks of a Genius
San Francisco: Berrett-Koehler, 2013

Review

Learning from Leonardo is a fascinating read and unveils much of Leonardo's unique personality, and especially the nature of his scientific and human genius. It seems conceptually to be the second volume of Capra's earlier book on Leonardo's science. I have a feel that these two books about Leonardo could in the future be considered as the most important works of Fritjof Capra, and they are certainly his highest achievements given the difficult nature of the subject, and the difficulties with translating and perusing an immense amount of data, which to this day will remain inaccessible to most humans on the globe. One probably needs to be a genius oneself to really penetrate into the universe of Leonardo.

The genius of Leonardo is so unique because it was so versatile. It cannot be compared with anything we know today, in a culture where specialization is required and where universal genius would be frowned upon as 'generalizing and imprecise.' Perhaps Leonardo had most in common with Aristotle, in that both men were general and precise at the same time, which is not achievable for most humans simply because of the sheer amount of data to process, ideas to develop, concepts to make, and hidden connections between seemingly separate subjects to make out and describe. I am well aware that in stating this, I made a comparison that limps as Aristotle was certainly not a great artist, nor did he excel in inventing and conceptualizing machines of any kind. It is then the unique combination, the unique synthesis of art, science and technology that makes the genius of Leonardo.

The main message of this book is that we, as a society, need to expand our understanding of the multi-faceted problems with an interdisciplinary perspective, rather than staying with the narrow focus of 'specialization' that modern science emphasizes so much. If we could see, as Leonardo did, the unity of all life, and recognize the fundamental interdependence of all natural phenomena, we would begin to design effective solutions to problems that we always thought were unsolvable. The author's intention was thus to outline the synthesis in thinking that Leonardo achieved 500 years ago. When you consider this enormous and enormously important claim, you cannot but think this must be a tremendously timely book.

What our science achieved only recently, over the last 30 years or so, within the framework of systems theory, was for Leonardo a natural and organic way of thinking, for at the core of Leonardo's synthesis was the understanding of living forms of nature. His conception of painting was scientific, too, in that it involved for him the study of natural forms in very minute detail, in a way that to my knowledge no other artist has ever undertaken. In this sense, the artist Leonardo and the scientist Leonardo cannot be separated: one part of his personality complemented the other. It is therefore important to understand both his art and his science, and then, as the author did in this book, arrive at a synthesis.

What only now emerges in modern science, namely an appreciation of the form and gestalt of matter, rather than its substance, Leonardo was lucidly aware of. He studied throughout his life the magic of water, its movements and flow nature and was by so doing a pioneer in the discipline known today as fluid dynamics. His manuscripts are filled with precise drawings of spiraling vortices. His studies seem to not have been appreciated by previous commenters, which makes Capra's contribution a very original one, as he delivers an in-depth analysis of Leonardo's 'water science,' and he based his analysis on extensive discussions with Ugo Piomelli, professor of fluid dynamics at Queen's University in Canada.

As Leonardo observed how water and rocks interact, he undertook ground-breaking studies in geology, even to the point that he identified folds of rock strata and outlining and evolutionary perspective 300 years before Charles Darwin.

In addition, Leonardo made extensive inquiries about plants. While this research was first intended as studies for paintings, it became so extensive that they resulted in genuine studies about the patterns of metabolism and growth that underlie all botanical forms. Other domains of study were mechanics, known today as statics, dynamics, and kinematics, thereby inventing a great number of machines. He also compared the way humans move their body and animals, by comparison, and what fascinated him most in this field of inquiry was the flight of birds. He became almost obsessed with flying, and thus designed highly original flying machines. But his science of flight, as Capra shows with his habitual systematic approach, involved numerous sub-disciplines such as aerodynamics, human and bird anatomy, and mechanical engineering.

Capra has great merit in identifying what he calls the 'grand unifying theme' in Leonardo's explorations of both the macrocosm and the microcosm, in order to gain an understanding of the nature of life. Capra reports that this quest reached its climax in the anatomical studies he carried out in Milan and Rome when he was over sixty, especially in his investigations of the human heart. Nobody had at that time an idea how the heart functions.

Finally as Leonardo approached old age, he became fascinated with the processes of reproduction and embryonic development. In his embryological studies, he described the life processes of the fetus in the womb in great detail.

I do not dare to utter any even slight critique of this enormous work by Fritjof Capra, and perhaps we should wait until the scientific world will eventually open their minds to this immensely enriching knowledge, until a real review of this book can be written. I will thus limit myself to providing, and commenting upon, some quotes from the book. My main focus shall be upon Fritjof Capra's discovery that Leonardo was a systems thinker before the scientific method was even officially implemented into modern science through the genius of Galileo, Bacon,

Kepler, Descartes and Newton. I believe that this discovery is so uncanny and important that it deserves closer scrutiny from the point of view of scientific methodology and epistemology. To begin with, Fritjof Capra writes in the Preface:

> Leonardo's view of natural phenomena is based partly on traditional Aristotelian and medieval ideas and partly on his independent and meticulous observations of nature. The result is a unique science of living forms and their continual movements, changes, and transformations—a science that is radically different from that of Galileo, Descartes, and Newton. /xi

In this context, it is important to see that the view that nature is organized like a machine, consisting of various assembled parts, was at no time a part of Leonardo's science:

> Unlike Descartes, Leonardo did not see the body as a machine, but he clearly recognized that the anatomies of animals and humans involve mechanical functions that can be appreciated only with an understanding of the basic principles of mechanics. / Id.

Fritjof Capra is convinced that if Galileo, Newton and their contemporaries had studied Leonardo's Notebooks they would possibly have come to different conclusions as to the organic nature of all life and the need for a systems view of life. He writes:

> Today, as we are developing a new systemic understanding of life with a strong emphasis on complexity, networks, and

patterns of organization, we are witnessing the gradual emergence of a science of qualities that has some striking similarities with Leonardo's science of living forms. /xii

It is very helpful in this context to consult the following 12 pages of the book that present a synopsis entitled 'Timeline of Scientific Discoveries.' This shows beyond any explicit description what an uncanny precursor the science of Leonardo was in the whole of human scientific history. In fact, much like Chinese science, it has forecasted most of our cherished scientific discoveries, thereby forcing publishers to revise some of their ingrained statements about the origins and proprietorship of scientific inventions.

I feel grateful and indebted to Fritjof Capra to have demonstrated with so much convincing detail all we collectively owe to the genius of Leonardo. This has not been revealed to me by what Hermann Grimm wrote nor has Vasari revealed this; it was probably beyond the horizon of historians, and it needed a modern scientist who has a systemic research perspective to lift the veil and tell the truth. He writes:

> One hundred years before Galileo Galilei and Francis Bacon, Leonardo single-handedly developed a new empirical approach to science, involving the systemic observation of nature, logical reasoning, and some mathematical formulations—the main characteristics of what is known today as the scientific method. /5

While Leonardo had begun his scientific investigations for writing a book about depicting natural phenomena in his drawings, it was according to Capra the fact that he carried these observations out in an 'organized, methodical fashion,' that makes us conclude that, while he looked at nature with the eyes of an artist, he became a scientist because of the intensity and systematic collection of his observations. The author writes:

> The systematic approach and careful attention to detail that Leonardo applied to his observations and experiments are characteristic of his entire method of scientific investigation. He would usually start from commonly accepted concepts and explanations, often summarizing what he had gathered from the classical texts before proceeding to verify it with his own observations. After testing the traditional ideas repeatedly with careful observations and experiments, Leonardo would adhere to tradition if he found no contradictory evidence; but if his observations told him otherwise he would not hesitate to formulate his own alternative explanations. /6

Now, to us today, one of the most striking features of human genius is the fact that a person gifted with genius is able to observe phenomena paralleled in entirely different fields of investigation, using a multi-disciplinary approach without perhaps being always conscious of this special multi-vectorial method of observation. Also, geniuses are known to work on many problems at the same time, profiting from every single insight in solving one, for solving the others. Capra writes:

Leonardo generally worked on several problems simultaneously and paid special attention to similarities of patterns in different areas of investigation. When he made progress in one area, he was always aware of the analogies and interconnecting patterns to phenomena in other areas, and would revise his theoretical ideas accordingly. This method led him to tackle many problems not just once but several times during different periods of his life, modifying his theories in successive steps as his scientific thought evolved over his lifetime. Leonardo's practice of repeatedly reassessing his theoretical ideas in various areas meant that he never saw any of his explanations as final. /Id.

Now in what ways, precisely, is Leonardo's science a science of organic patterns, and why can we say today that he was one of the first systems thinkers in the history of science? Let us first look at what systemic thinking is all about. Capra develops this on pages 7 to 9 of the book, still within the Prologue. Capra explains that throughout the history of Western science, there was a conceptual dilemma between the parts and the whole. He continues:

> The emphasis on the parts has been called mechanistic, reductionist, or atomistic; the emphasis on the whole holistic, organismic, or ecological. In twentieth-century science, the holistic perspective has become known as 'systemic' and the way of thinking it implies as 'systemic thinking.' /7

But historically we can trace this dichotomy back to the ancient Greek who distinguished between pattern and substance (matter). Capra further explains:

> These two very different lines of investigation have been in competition with one another throughout our scientific and philosophical tradition. The study of matter was championed by Democritus, Galileo, Descartes, and Newton; the study of form by Pythagoras, Aristotle, Kant, and Goethe. Leonardo clearly followed the tradition of Pythagoras and Aristotle in developing his science of living forms, their patterns of organization, and their processes of growth and transformation. Indeed, systemic thinking lies at the very core of his approach to scientific knowledge. (...) Nature as a whole was alive for Leonardo. He saw the patterns and processes in the microcosm as being similar to those in the macrocosm. /8

It is here where Leonardo's special and uncanny drawing talent came in for it enabled him to *depict* rather than *describe* their shapes, ... 'and he analyzed them in terms of their proportions rather than measured quantities.' /Id.

Capra emphasizes that Leonardo's approach to nature was dynamic, not static. For example he portrays the forms of nature, be it mountains, rivers, plants, or the human body, as in ceaseless movement and transformation.

> He studies the multiple ways in which rocks and mountains are shaped by turbulent flows of water, and how the organic forms of plants, animals, and the human body are shaped by their metabolism. The world Leonardo portrays, both in his art and his science, is a world in development and flux, in which all configurations and forms are merely stages in a continual process of transformation. /8-9

The last sub-chapter of the Prologue of the book—on which I only comment in this review (for the book is so immensely rich in information that it's impossible to review it in further detail) is entitled 'Inspiration for Our Time.' The author writes:

> The great challenge of our time is to build and nurture sustainable communities—communities designed in such a way that their ways of life, businesses, economy, physical structures, and technologies respect, honor, and cooperate with nature's inherent ability to sustain life. The first step in this endeavor, naturally, must be to understand how nature sustains life. It turns out that this involves a new ecological understanding of life, also known as 'ecological literacy,' as well as the ability to think systemically—in terms of relationships, patterns, and context. /9

> This new science is being formulated in a language quite different from Leonardo's. As we shall see throughout this book, however, the underlying conception of the living world as being fundamentally interconnected, highly complex, creative, and imbued with cognitive intelligence is quite similar to Leonardo's vision. /10

To repeat it, I consider Capra's statement of Leonardo having been the first systems thinker in the modern sense of the word as the work hypothesis of this book—as it was already the point of departure of his previous book, *The Science of Leonardo (2007)*. The book in its further elaboration is conceptualized as a detailed analysis of Leonardo's notebooks with the focus on confirming and corroborating the work hypothesis with tangible facts. This way of proceeding is methodologically simple and convincing. The author further sharpens the focus of investigation in tracing the development, from the 17th century to today, how the mechanistic worldview developed definitely into a holistic science that, as astonishing as it may sound, connects us back to the science of Leonardo, and even the worldview of the ancient philosophers Pythagoras and Heraclitus.

> At the core of the new understanding of life is a shift of metaphors from seeing the world as a machine to understanding it as a network. Exploring this shift without prejudice, driven by intellectual curiosity, will be beneficial in many ways. Individually, it will help us to better deal with our health, seeing our organism as a network of components with both physical and cognitive/emotional dimensions. As a society, the exploration of networks will help us to build a sustainable future, grounded in the awareness of ecological networks and the interconnectedness of our major problems. Such exploration will also help us manage our organizations, which are social networks of increasing complexity. /Id.

Quotes

- One hundred years before Galileo Galilei and Francis Bacon, Leonardo single-handedly developed a new empirical approach to science, involving the systematic observation of nature, logical reasoning, and some mathematical formulations—the main characteristic of what is known today as the scientific method. /5

- The empirical approach came naturally to Leonardo. He was gifted with exceptional powers of observation, which were complimented by great drawing skills. He was able to draw the complex swirls of turbulent water or the swift movements of a bird in flight with a precision that would not be reached again until the invention of serial photography. /Id.

- What turned Leonardo from an artist with exceptional gifts of observation into a scientist was his recognition that his observations, in order to be scientific, needed to be carried out in an organized, methodical fashion. /Id.

- The systematic approach and careful attention to detail that Leonardo applied to his observations and experiments are characteristic of his entire method of scientific investigation. He would usually start from commonly accepted concepts and explanations, often summarizing what he had gathered from the classical texts before proceeding to verify it with his own observations. After testing the traditional ideas repeatedly with careful observations and experiments, Leonardo would adhere to tradition if he found no contradictory evidence; but if his observations told him otherwise he would not hesitate to formulate his own alternative explanations. /6

- When he made progress in one area, he was always aware of the analogies and interconnecting patterns to phenomena in other areas, and would revise his theoretical ideas accordingly. This method led him to tackle many problems not just once but several times during different periods of his life, modifying his theories in successive steps as his scientific thought evolved over his lifetime. /Id.

- Leonardo's practice of repeatedly reassessing his theoretical ideas in various areas meant that he never saw any of his explanations as final. /Id.

- This technique of using simplified theoretical models to understand complex phenomena put him centuries ahead of his time. /7

- Ever since the days of early Greek philosophy, there has been this tension between substance and pattern, between matter and form. /Id.

- These two very different lines of investigation have been in competition with one another throughout our scientific and philosophical tradition. The study of matter was championed by Democritus, Galileo, Descartes, and Newton; the study of form by Pythagoras, Aristotle, Kant, and Goethe. Leonardo clearly followed the tradition of Pythagoras and Aristotle in developing his science of living forms, their patterns of organization, and their processes of growth and transformation. Indeed, systemic thinking lies at the very core of his approach to scientific knowledge. /8

- Nature as a whole was alive for Leonardo. He saw the patterns and processes in the microcosm as being similar to those in the macrocosm. /Id.

- He preferred to *depict* the forms of nature rather than *describe* their shapes, and he analyzed them in terms of their proportions rather than measured quantities. /Id.

- He studies the multiple ways in which rocks and mountains are shaped by turbulent flows of water, and how the organic forms of plants, animals, and the human body are shaped by their metabolism. /9

- The world Leonardo portrays, both in his art and in his science, is a world in development and flux, in which all configurations and forms are merely stages in a continual process of transformation. /Id.

- The great challenge of our time is to build and nurture sustainable communities—communities designed in such a way that their ways of life, businesses, economy, physical structures, and technologies respect, honor and cooperate with nature's inherent ability to sustain life. The first step in this endeavor, naturally, must be to understand how nature sustains life. It turns out that his involves a new ecological understanding of life, also known as 'ecological literacy,' as well as the ability to think systemically—in terms of relationships, patterns, and context. /Id.

- Contemporary science no longer sees the universe as a machine composed of elementary building blocks. We have discovered that the material world, ultimately, is a network of inseparable patterns of relationships; the planet as a whole is a living, self-regulating system. /Id.

- Leonardo's view of the essential role of water in biological life is fully borne out by modern science. Today we know not only that all living organisms need water for transporting nutrients to their tissues but also that life on Earth began in water. The first living cells originated in the primeval oceans more than three billion years ago, and ever since that time all the cells that compose living organisms have continued to flourish and evolve in watery environments. Leonardo was completely correct in viewing water as the carrier and matrix of life. /18

- Another reason Leonardo was so fascinated by water is that he associated it with the fluid and dynamic nature of organic forms. Ever since antiquity, philosophers and scientists had recognized that biological form is more than shape, more than a static configuration of components in a whole. There is a continual flux of matter through a living organism, while its form is maintained; there is growth and decay, regeneration and development. This dynamic conception of living nature is one of the main themes in Leonardo's science and art. /21

- The process of metabolism, the hallmark of biological life, involves a continual flow of energy and matter through a living organism—the intake and digestion of nutrients and the excretion of waste products—while its form is maintained. Thus, metaphorically, one could visualize a living organism as a whirlpool, even though the metabolic processes at work are not mechanical but chemical. /22

- In Leonardo's time, the scientific study of flow phenomena, now known as fluid dynamics, was entirely new. It was a field of study he himself created single-handedly. /32

- This method of using simplified models to analyze the essential features of complex phenomena is an outstanding characteristic of our modern scientific method. The fact that Leonardo used it repeatedly is truly remarkable. /33

- Realizing that the mathematics of his time was inappropriate for describing the ceaseless movements and transformations of flowing water, and that his own geometry of transformations was too rudimentary for modeling complex flow phenomena, Leonardo chose a third option. Instead of mathematics, he used his exceptional facility of drawing to

document his observations in pictures that can be strikingly
beautiful while at the same time playing the role of
mathematical diagrams. /43

- Leonardo's conception of the Earth as being alive,
  manifesting patterns and processes common to all living
  systems, was a forerunner of the modern Gaia theory, which
  views our planet as a living, self-organizing, and
  self-regulating system. /68

- All of Leonardo's science is utterly dynamic. He portrays
  nature's forms in ceaseless movement and transformation,
  recognizing that living forms are continually shaped and
  transformed by underlying processes. This dynamic
  conception of nature is evident in his studies of anatomy,
  botany, and fluid dynamics, and it is perhaps most tricking
  in his geology. /Id.

- Understanding a phenomenon, for him, always meant
  connecting it with other phenomena through a similarity of
  patterns. That he was able to associate the annual rings in
  the branches of trees with the growth rings in the horns of
  sheep is remarkable enough. To use the same analysis to
  infer the lifespan of a fossilized shell is extraordinary. /84

- The transition of Leonardo's botanical drawings from
  studies for paintings to scientific illustrations was
  accompanied by a series of texts that represent his first
  purely scientific inquiries into the nature of botanical forms
  and processes. To appreciate the significance of his evolution
  in Leonardo's thought we first need to have some idea of the
  history of botany since antiquity, which formed the
  intellectual context within which he operated. /107

- In contrast to his contemporaries, Leonardo not only
  depicted plants accurately but also sought to understand the
  forces and processes underlying their forms. In these studies,
  often based on observations that were astonishing for their
  time, he pioneered the emergence of botany as a genuine
  science. /112

- In his study of plants and animals, Leonardo identified the
  soul as the vital force underlying their formation and
  growth. Following Aristotle, he conceived of the soul as
  being built up in successive levels, corresponding to levels of

organic life. The first level is the 'vegetative soul,' which controls the organism's metabolic processes. The soul of plants is restricted to this metabolic level of a vital force. The next higher form is the 'animal soul,' characterized by autonomous motion in space and by feelings of pleasure and pain. The 'human soul,' finally, includes the vegetable and animal souls, but its main characteristic is reason. /153

- In view of Leonardo's brilliant achievements in mechanical engineering and his extensive applications of principles of mechanics to the body's 'mechanical instruments,' it is tempting—but, in my view, erroneous—to believe that Leonardo saw the entire human body as a machine. Many Leonardo scholars have, implicitly or explicitly, taken this view. /155

- It amounts to projecting a reductionist, mechanistic model of the human body onto Leonardo's scientific views of the macro- and microcosm—views that were utterly organic and unmarried by the mind-body split introduced by Descartes more than one hundred years after Leonardo's death. /Id.

- He was a mechanical genius who invented countless machines and mechanical devices, and he maintained a lively interest in the theory of mechanics during most of his mature life. Yet his science as a whole was not mechanistic. /157

- The grand unifying theme of Leonardo's explorations of the macro- and microcosm was his persistent quest to understand the nature of life. Over the years, as he studied, drew, and painted the flows of water and air, the rocks and sediments of the Earth, the growth patterns of plants, and the anatomy of the human body, he correctly identified several of life's key biological characteristics. /281

- Nature as a whole was live for Leonardo, and he saw similar patterns and processes in both the macrocosm of the living Earth and the microcosm of an individual organism. In view of this systemic approach—seeking to understand a natural phenomenon by linking it to other phenomena through a similarity of patterns—it is not surprising that Leonardo developed a conception of life that was deeply ecological. /Id.

Chapter Thirteen
The Systems View of Life

The Systems View of Life

A Unifying Vision
With Pier Luigi Luisi
Cambridge: Cambridge University Press, 2014

Review

The Systems View of Life is Capra's latest publication and it summarizes and condenses the *summa totus* of his lifetime achievements. It also reflects the systems research of his co-author, Pier Luigi Luisi, who does not agree in all the points with Fritjof Capra. It is a textbook targeting an audience of both college students and professionals.

Regarding the form of the book, allow me to note that I was disenchanted with the small font size which makes reading the book really not easy. As the publisher used glossy paper, the book is also unusually heavy.

While it is difficult to review this book because of its extensive coverage of all subjects that can possibly be relevant to the systems view of life, I will stick to my method of commenting on some of the quotes taken from the book while reading it.

Let me first communicate my general impression of the book. It is one of the best of its kind, and really complete in terms of content.

Contrary to most other systems researchers, Fritjof Capra and his co-authors always write in a style and diction every intelligent lay reader can understand. Of course, as the book is meant as a textbook, it contains much of the content of Capra's previous books on systems theory, especially *The Web of Life* (1997) and *The Hidden Connections* (2002). But that is definitely not a disadvantage: we are facing here different target audiences and we have an intention of conclusiveness and completeness.

We have systems theory around since more than thirty years, yet to this day, a textbook that covers the matter in all its facets had been lacking. Systems thinking, while it is a development of the 20th century within the framework of modern science is not something new. All ancient scientific traditions were holistic and systemic, to mention only Chinese science and traditional medicine, and the science traditions of Egypt, Persia, and India.

The emergence of systems thinking within our science tradition was delayed because of the focus upon *hierarchies*, so typical for patriarchy, together with the fact that in the late Renaissance the scientific method became associated with a mechanistic approach to nature, the human body, and the cosmos at large. While Fritjof Capra, in two of his recent books, showed that Leonardo da Vinci, who was the first modern scientist, was a systems thinker and had an organic view of nature, he also explained in these books, and in the present book, how with Bacon, Galilei, Newton, and Descartes, the mechanistic orientation of our modern science paradigm was written in stone.

The achievement of systems thinking, to mention especially Ludwig von Bertalanffy, Ilya Prigogine, Humberto Maturana and Francisco Varela, was to shift the scientific focus from hierarchies to networks, and from principles to patterns of organization.

The present volume integrates the ideas, models, and theories underlying what Fritjof Capra coined as 'the systems view of life' into a coherent framework. The authors more explicitly explain:

> Taking a broad sweep through history and across scientific disciplines, the authors examine the appearance of key concepts such as autopoiesis, dissipative structures, social networks, and a systemic understanding of evolution. The implications of the systems view of life for healthcare, management, and our global ecological and economic crises are also discussed.

The *Introduction* is highly useful for clarifying basic notions that are recurring in the book, such as the *scientific method,* the notion of a *paradigm,* or *paradigm shift,* as well as the terms *mechanism, holism,* and *deep ecology.*

We are reminded in this introductory chapter that in ancient times, both in the West and the East, a holistic and dynamic paradigm reigned in what was called, still at the time of Newton not science, but 'natural philosophy.'

While in ancient Greece, a philosophical school, called the Milesians made no distinction between animate and inanimate nature, nor between spirit and matter, in ancient China, the dynamic comprehension of life was an outflow of the insight in the nature of *Tao.* This was not just one philosophical school but a general paradigm, which also reigned strong in native civilizations around the world:

> The ancient Chinese philosophers believed that the ultimate reality, which underlies and unifies the multiple phenomena

we observe, is intrinsically dynamic. They called it *Tao*—the way, or process, of the universe. For the Taoist sages all things, whether animate or inanimate, were embedded in the continuous flow and change of the *Tao*. The belief that everything in the universe is imbued with life has also been characteristic of indigenous spiritual traditions throughout the ages. /1

I shall *not* review the book as very much of it is just a repetition of the earlier books, however perhaps put in a more systematical style. I believe the book cannot really be reviewed, as it's a textbook, and a review can never reflect all the high amount of detail of the original.

I shall therefore publish the quotes only, for they represent substantial lecture in themselves, explaining all the main facets and details of the 'systems view' comprehensively.

Quotes

- The particles of light were first called 'quanta' by Einstein—hence the term 'quantum theory'—and are now known as photons. /71

- An electron is neither a particle nor a wave, but it may show particle-like aspects in some situations and wave-like aspects in others. While it acts like a particle, it is capable of developing its wave nature at the expense of its particle nature, and vice versa, thus undergoing continual transformations from participle to wave and from wave to particle. This means that neither the electron nor any other atomic 'object' has any intrinsic properties independent of its environment. /Id.

- In this penetration into the world of submicroscopic dimensions, a decisive point is reached in the study of atomic nuclei, in which the velocities of protons and neutrons are often so high that they come close to the speed of light. This fact is crucial for the description of their interactions, because any description of natural phenomena involving such high velocities has to take the theory of relativity into account. To understand the properties and interactions of subatomic particles we need a framework that incorporates not only quantum theory but also relativity theory; and it is relativity theory that reveals the dynamic nature of matter to its fullest extent. /75-76

- The application of complexity theory, technically known as nonlinear dynamics, raised systems thinking to an entirely new level and provided the conceptual basis for vastly more sophisticated formulations of the systems view of life. /97

- Most of nature is very, very complicated. How could one describe a cloud? A cloud is not a sphere ... It is like a ball but very irregular. A mountain? A mountain is not a cone ... If you want to speak of clouds, of mountains, or rivers, of lightning, the geometric language of school is inadequate. /Id.

- Of all those, the fractal patterns of clouds, which originally inspired Mandelbrot to search for a new mathematical language, are perhaps the most stunning. Their self-similarity stretches over seven orders of magnitude, meaning that the border of a cloud magnified 10 million times still shows the same familiar shape. /119

- 'Autopoiesis' is a term coined by Maturana and Varela in the 1970s. *Auto*, of course, means 'self' and refers to the autonomy of self-organizing systems; and *poiesis* (which shares the same Greek root as the word 'poetry') means 'making.' So, autopoiesis means 'self-making.' /129

- According to Maturana and Varela, the main characteristic of life is self-maintenance due to the internal networking of a chemical system that continuously reproduces itself within a boundary of its own making. /Id.

- Where is cellular life localized? Is there a particular reaction, a particular magical spot, where we can put a tag to say: here is life? There is an obvious and very important answer to this question: life is not localized; life is a global property, arising from the collective interactions of the molecular species in the cell. /132

- Where is the life of an elephant, or of a given person, localized? Again, there is no localization; the life of any large mammal is the organized, integrated interaction of heart, kidneys, lungs, brain, arteries and veins. And each of these organs, which are connected in a network, can be seen in turn as a network of several different tissues and specialized organelles; and each tissue and each organelle is the networking of many different kinds of cells. /133

- Emergence, in the most classic interpretation, means in fact the arising of novel properties in an ensemble, novel in the sense that they are not present in the constituent parts. /Id.

- The properties of life are emergent properties which cannot be reduced to the properties of its components. /Id.

- The cell, like any living organism, does not need any information from the environment to be itself: all information needed for a fly to be a fly is contained inside the fly, and the same is true for the elephant. In the language of epistemology, we say that the cell, and by inference every living organism, is an operationally closed system. /133

- According to Maturana and Varela ... the organism interacts with the environment in a 'cognitive' way whereby the organism 'creates' its own environment and the environment permits the actualization of the organism. /134

- Again, what is valid for cellular life can be considered valid for any form of life. The primary literature distinguishes between first-order and second-order (multicellular) autopoietic systems. Thus, an organ like the heart can be seen as an autopoietic system, as it is capable of self-sustainment through a series of processes which regenerate all components within its own boundary. On the other hand, this complex autopoietic system is composed of smaller autopoietic units, down to the single cells of various kinds; and the entire human being can also be seen as an autopoietic system. For us, here it is important to see the relation to the systems view of life; we can now say that life, more precisely, may be seen as a system of interlocked autopoietic systems. /135

- The relation between autopoiesis, operational closure, circular logic, and biological autonomy is also important. Autopoiesis is the particular self-organization of life that specifies the processes which, within a circular logic, permit the regeneration of the components. The notion of biological autonomy then also means that the living organism is an operationally closed system with a circular logic. /Id.

- The notion of structural determinism sheds new light on the age-old philosophical debate about freedom and determinism. According to Maturana, the behavior of a living organism is determined. However, rather than being determined by outside forces, it is determined by the organism's own structure—a structure formed by a succession of autonomous structured changes. Hence, the behavior of the living organism is both determined and free. /136

- From the generalizations emerges the important insight that social networks exhibit the same general principles as biological networks. There is an organized ensemble with internal rules that generates both the network itself and its boundary (a physical boundary in biological networks, and a cultural boundary in social networks). Each social system—a political party, a business organization, a city, or a school—is characterized by the need to sustain itself in a stable but dynamic mode, permitting new members, materials, or ideas to enter the structure and become part of the system. These newly entered elements will generally be transformed by the internal organization (i.e., the rules) of the system. /137

- [F]or the entire realm of the biological world, the equivalence between autopoiesis and life holds true, and we may safely adopt this generalization. This means that, in order to determine whether a system is living or not, it will be sufficient to see whether it is autopoietic. /138

- What then are the criteria of autopoiesis? Generally, the simplest criterion is to see whether the system is capable of sustaining itself due to self-generating processes taking place within its boundary, the boundary being of its own making. Take a cell: it satisfies these criteria, but a virus does not. /Id.

- It is important to emphasize that autopoiesis does not indicate reproduction as a criterion for life. According to the main philosophy of the Santiago school, reproduction is a property of life than can be present, or not, depending upon conditions. Of course, nobody denies that reproduction is the main process for biodiversity and life's unfolding on Earth, but the point should be made that, before talking about reproduction, one must have a 'container' and the pattern of self-organization to make reproduction possible, and autopoiesis is this preliminary setup. Within that, one can have reproduction as one of the kinetic modes of the autopoietic being, or not (for example, when the organism is in homeostasis, or sterile, or does not need to reproduce). Take a colony of bacteria which have lost the capability of reproduction, but are self-sustaining and are provided with a normal metabolism. Would one not consider this colony living? Or, take babies or old people who cannot reproduce—are they not living? / 138-139

- In molecular and structural terms, we can describe death as the disintegration of the autopoietic organization that characterizes life. The essence of life is integration: namely, the linking of the various organs—heart and kidneys, brain and lungs, etc.—with one another. When this mutual linkage disappears, the system is no longer an integrated unity and death occurs. / 139

- Thus, death, seen in this way, is a progressive process, and corresponds to the destruction of the emergent properties of the various levels characterizing the complexity of the entire organism. / Id.

- The emphasis and overall concern here is not to define cognition in terms of an input from the external world acting on the perceiver, but rather to explain cognition and perception in terms of the internal structure of the organism. / 141

- Cognition, then, operates at various levels, and as the sophistication of the organism grows, so does its sensorium for the environment, and so does the extent of co-emergence between organism and environment. Thus we got from unicellular to multicellular organisms, where we can have flagella and light-or sugar-sensitive receptors, to the development of sensitive tentacles in the first aquatic organisms, and up to the higher cognitive functions in fish. In all these cases, the organisms contribute to the 'creation' of their environments. / Id.

- Mind is always present in a bodily structure; and, vice versa, a truly living organism must be capable of cognition (the process of knowing). The same holds for human consciousness. Consciousness is not a transcendent entity, but it is always manifest within an organic living structure. /142

- The term 'environment' can represent quite different things, depending on the levels of life we consider: it can be the milieu in which cells swim, or the habitat where animals live, or the urban environment or humans. In all cases, as in the case of the bio-logic of life, there is a conceptual similarity: the interaction between the living organism and the environment is a dynamic one based on co-emergence, where the living organism and the environment become one through cognitive interactions. /143

- Another important clarification we need to make right away concerns the difference between static and dynamic aspects of self-organization. Reactions or processes under thermodynamic control usually lead to a final equilibrium situation where all is still, in the sense that the relative concentrations do not change anymore. Self-organization is, however, important also in dynamic systems that operate far from equilibrium. /144

- The phenomenon of emergence must be considered together with self-organization. The term 'emergence' refers to the arising of novel properties of the organized structure, novel in the sense that they are not present in the parts or components. /145

- However, the notion of emergence goes beyond life at the individual level, extending to living colonies and, in general, to social life. Indeed, the structures created by social insects—beehives, anthills, and so on—emerge at the social level. In a beehive, for example, each bee appears to behave as an independent element, acting apparently in its own account, but the whole population of bees produces a highly sophisticated structure emerging from their collective activities. /157

- The view that the emergent properties of molecules are not explicable as a matter of principle on the basis of the components is opposed by several scientists, who argue that

> this is tantamount to assuming a mysterious force of some undefined nature—a kind of vitalistic principle. The systems view of life takes a third position, asserting that there is no need to assume any mysterious force to explain emergent properties, but that the focus on relationships, patterns, and underlying processes is essential. Once this is accepted, practical difficulties will still be relevant, and the distinction between strong and weak emergence may not always make sense. /157

- In classical thermodynamics, the dissipation of energy in heat transfer, friction, etc., is always associated with waste. Prigogine's concept of a dissipative structure introduced a radical change in this view by showing that in open systems dissipation becomes a source of order. /159

- According to Prigogine's theory, dissipative structures not only maintain themselves in a stable state far from equilibrium but may also even evolve. When the flow of energy and matter through them increases, they may go through new instabilities and transform themselves into new emergent structures of increased complexity. In the language of nonlinear dynamics, the system encounters bifurcation points at which it may branch off into entirely new states, each characterized by a specific attractor, where new structures and new forms of order emerge. /Id.

- One of Prigogine's greatest achievements has been to resolve the paradox of the two contradictory views of evolution in physics and biology—one of an engine running down, and the other of a living world unfolding toward increasing order and complexity. In Prigogine's theory, the second law of thermodynamics (the law of ever-increasing entropy, or disorder) is still valid, but the relationship between entropy and disorder is seen in a new light. At bifurcation points, states of greater order may emerge spontaneously without contradicting the second law of thermodynamics. The total entropy of the system keeps increasing, but this increase in entropy is not a uniform increase in disorder. In the living world, order and disorder are always created simultaneously. /159-160

- Emergence is one of the hallmarks of life. It has been recognized as the dynamic origin of development, learning, and evolution. In other words, creativity—the generation of new forms—is a key property of all living systems. And

- since emergence is an integral part of the dynamics of open systems, open systems develop and evolve. Life constantly reaches out into novelty. /160-161

- The outstanding feature of these feedback loops is that they link together living and nonliving systems. We can no longer think of rocks, animals, and plants as being separate. Gaia theory shows that there is a tight interlocking between the planet's living parts—plants, microorganisms, animals—and its nonliving parts—rocks, oceans, and the atmosphere. /164

- The golden spiral is a particular logarithmic spiral that grows by a factor of Φ (the golden ratio) for every quarter turn. /176

- Instead of being a machine, nature at large turns out to be more like human nature—unpredictable, sensitive to the surrounding world, and influenced by small fluctuations. Accordingly, the appropriate way to approaching nature to learn about her complexity and beauty is not through domination and control but through respect, cooperation, and dialogue. /180

- In fact, the foundation of Darwinism is the idea that we all come from a common ancestor with modifications, and that is tantamount to saying that all living forms, from trees to fish, and from mammals to birds—since they all come from the same primordial ancestor—are linked to each other by a network of parenthood. /182

- These and other observations have led systems biologists to an understanding of evolution that is considerably richer and more diverse than the modern synthesis. According to this new systemic understanding, the unfolding of life on Earth proceeded through three major avenues of evolution. The first, but perhaps least important, is the random mutation of genes, the centerpiece of the neo-Darwinian theory. These gene mutations, caused by chance errors in the self-replication of DNA, do not seem to occur frequently enough to explain the evolution of the great diversity of life forms, given the well-known fact that most mutations are harmful and very few result in useful variations. /193

- Thus, through the evolutionary process a rich biodiversity appeared on our planet. The study of the relationships between genome structures across different biological species is a fascinating new discipline, known as comparative genomics. For example, limiting the analysis to our human species, we have discovered that more than 99% of our genes have a related copy in the mouse—despite more than 500 million years of evolutionary separation. Moreover, the differences between human races worldwide are thought to be coded by only 0.1% of the human genome. This means that molecular genetics has demonstrated that there are no significant differences among the various human races. /Id.

- The ensemble of genes in the human organism consists of a sequence of 3 billion pairs of bases, and each of them has been identified. The human genome has captured the mass media, and is even seen as 'the book of life.' However, ... this notion is very problematic. There are in the human genome around 25,000 genes, and since human life is based on a considerably larger number of proteins, the classic notion of the 'central dogma of molecular biology'—one gene/one protein—does not hold true anymore. /195

- Darwinism actually had nothing to do with these ideas of 'social Darwinism,' which are an extrapolation based on the false assumption that the evolution of human society proceeds according to the biological evolution of simple organisms, and on the additional false assumption that all that is 'natural'—seen in the scenario of life in nature—should be right from the moral point of view. However, even if we discard social Darwinism and the conservative aspects of sociobiology, the fact remains that natural selection, generally, can be seen as some sort of competition among different living groups. /204

- Yes, mutations are random, and so is gene mixing, but evolution as a whole does not proceed randomly at all. Nature is very choosy in selecting a viable mutation. In order to be 'accepted,' the mutation has to respect several conditions and constraints. First of all, there is the principle of structural determinism, which implies that only those changes can be accepted that are consistent with the existing inner structure and organization of the living organism. Moreover, the mutation must produce a minimal perturbation so as to respect the main function of the living cell, which is self-maintenance. The cell's proper individuality must be preserved: a liver cell tends to remain a liver cell, a nerve cell remains a nerve cell; and so on. In

> addition, the mutation must permit adaptations to environmental changes; and finally, it has to comply with the laws of physics and chemistry that govern the cell's metabolism. /214

> In the view of evolution we have presented here, there is always a subtle interplay between contingency and determinism. Although chance events trigger evolutionary changes, the emerging new forms of life are not the result of these chance events alone, but of a complex, nonlinear dynamics—a web of factors involving not only genetics, but also the constraints of the physical structure of the organism and its context, as well as the ever-changing environment. /214-215

> Contemporary studies in primatology have completely confirmed Darwin's revolutionary view. Today we know that the genomes of chimpanzees and humans differ only by a mere 1.6%. ... In addition, it is well known that much of the chimpanzee facial repertoire is similar to our own. /246

> In other words, violence is not a general human characteristic, but rather a specifically male human characteristic. /247

> Bateson listed a set of criteria that systems have to satisfy for mind to occur. Any system that satisfies those criteria will be able to develop the processes we associate with mind—learning, memory, decision-making, and so on. In Bateson's view, these mental processes are a necessary and inevitable consequence of a certain complexity that begins long before organisms develop brains and higher nervous systems. He also emphasized that mind is manifest not only in individual organisms but also in social systems and ecosystems. /253

> In other words, cognition is the very process of life. The organizing activity of living systems, at all levels of life, is mental activity. The interactions of a living organism—plant, animal, or human—with its environment are cognitive interactions. Thus life and cognition are inseparably connected. Mind—or, more accurately, mental activity—is immanent in matter at all levels of life. /254

> Maturana wrote in his paper 'Biology of Cognition: 'Living systems are cognitive systems, and living as a process is a

> process of cognition. This statement is valid for all organisms, with and without a nervous system. (Id.)

- Our cognitive process differs from the cognitive processes of other organisms only in the kinds of interactions into which we can enter, such as linguistic interaction, and not in the nature of the cognitive process itself. /255

- In other words, a living system has the autonomy to decide what it is to notice and what will disturb it. This is the key to the Santiago theory of cognition. The structural changes in the system constitute acts of cognition. By specifying which perturbations from the environment trigger changes, the system specifies the extent of its cognitive domain; it 'brings forth the world,' as Maturana and Varela put it. /256

- While the conceptual boundaries between soul and spirit were often fluctuating in the philosophical schools of antiquity, both soul and spirit were described in the languages of ancient times with the metaphor of the breath of life. The words for 'soul' in Sanskrit (atman), Greek (psyche), and Latin (anima) all mean 'breath.' The same is true of the words for 'spirit' in Latin (spiritus), Greek (pneuma) and Hebrew (ruah). These, too, mean breath. /Id.

- The relationship between mind and brain, therefore, is one between process and structure. Moreover, the brain is not the only structure through which the process of cognition operates. The entire structure of the organism participates in the process of cognition, whether or not the organism has a brain and a higher nervous system. /257

- Consciousness—that is, conscious, lived experience—unfolds at certain levels of cognitive complexity that require a brain and a higher nervous system. In other words, consciousness is a special kind of cognitive process that emerges when cognition reaches a certain level of complexity. /Id.

- Reflective consciousness involves a level of cognitive abstraction that includes the ability to hold mental images, allowing us to formulate values, beliefs, goals, and strategies. This evolutionary stage established a fundamental link between consciousness and social phenomena, because with the evolution of language arose not only the inner

> world of concepts and ideas but also the social world of organized relationships and culture. /260

- And more recently, the Santiago theory of cognition has made it clear that cognition itself is not a representation of an independently existing world, but rather a 'bringing forth' of a world through the process of living. /262

- As far as first-person experience is convened, three main approaches are being pursued. The adherents of all three insist that they are not talking about a casual inspection of experience but about using strict methodologies that require special skills and sustained training, just like the methodologies of other areas of scientific observation. The first approach is introspection, a method developed at the very beginning of scientific psychology. The second is the phenomenological approach in the strict sense, as developed by Husserl and his followers. The third approach consists of using the wealth of evidence gathered from meditative practice in various spiritual traditions. /264

- Emotions are complex patterns of chemical and neural responses that have specific regulatory functions. Most emotional responses have a long evolutionary history; they automatically provide organisms with survival-oriented behaviors. /269

- Communication, according to Maturana, is not primarily a transmission of information, but rather a coordination of behavior between living organisms. Such mutual coordination of behavior is the key characteristics of communication for all living organisms, with or without nervous systems, and it becomes more and more subtle and elaborate with nervous systems of increasing complexity. /270

- Language arises when a level of abstraction is reached at which there is symbolic communication. This means that we use symbols—words, gestures, and other signs—as effective tools for the mutual coordination of our actions. In this process, the symbols become associated with abstract mental images of objects. The ability to form such mental images turns out to be a crucial characteristic of reflective consciousness. Abstract mental images are the basis of concepts, values, goals, and strategies. /Id.

- Spiritual experience is an experience of aliveness of mind and body as a unity. /277

- Spiritual experience—the direct, nonintellectual experience of reality in moments of heightened aliveness—is known as a mystical experience because it is an encounter with mystery. Spiritual teachers throughout the ages have insisted that the experience of a profound sense of connectedness, of belonging to the cosmos as a whole, which is the central characteristic of mystical experience, is ineffable—incapable of being adequately expressed in words or concepts—and they often describe it as being accompanied by a deep sense of awe and wonder together with a feeling of great humility. /278

- Spirituality is a way of being grounded in a certain experience of reality that is independent of cultural and historical contexts. Religion is the organized attempt to understand spiritual experience, to interpret it with words and concepts, and to use this interpretation as the source of moral guidelines for the religious community. /280

- The original purpose of religious communities was to provide opportunities for their members to relive the mystical experiences of the religion's founders. For this purpose, religious leaders designed special rituals within their historical and cultural contexts. These rituals might involve special places, robes, music, psychedelic drugs, and various ritualistic objects. In many religions, these special means to facilitate mystical experience become closely associated with the religion itself and are considered sacred. /282

- A further important similarity between the ways of the physicist and the mystic is the fact that their observations take place in realms that are inaccessible to the ordinary senses. In modern physics, these are the realms of the atomic and subatomic world; in mysticism, they are nonordinary states of consciousness in which the everyday sensory work is transcended. In both cases, access to these nonordinary levels of experience is possible only after long years of training within a rigorous discipline, and in both fields the 'experts' assert that their observations often defy expressions in ordinary language. /286

- Buddhists apply their conception of phenomena in terms of processes and relationships also to the structures of the mind. In accordance with the principles of emptiness, they hold that there is no independently existing, immutable self. Rather, the self is an emergent property that changes from moment to moment—a notion dear to modern cognitive scientists. /290

- Rather than judging unethical behavior as bad in an absolute sense, Buddhists consider it 'unskillful,' because it is a hindrance to one's self-realization. /Id.

- Ecological literacy involves not only the intellectual understanding of the basic principles of ecology but also the deep ecological awareness of the fundamental independence of all phenomena and of the fact that, as individuals and societies, we are embedded in, and dependent upon, the cyclical processes of nature. And since this awareness, ultimately, is grounded in spiritual awareness, it is evident that ecological literacy has an important spiritual dimension. /291

Bibliography
German and French Editions of Fritjof Capra's Books

Das Neue Denken
Die Entstehung eines ganzheitlichen Weltbildes im Spannungsfeld
zwischen Naturwissenschaft und Mystic
Frankfurt/M.: Fischer Verlag, 2015
Erste Auflage 1992

Die Capra Synthese
Grundlegende Texte des führenden Interpreten ganzheitlichen
Forschens und Denkens
Frankfurt/M.: Fischer-Scherz Verlag, 1998

Synthese
Neue Bausteine für das Weltbild von Morgen
München: Droemer/Knaur, 2000

Wendezeit
Bausteine für ein neues Weltbild
München: Droemer Knaur, 2004

Wendezeit im Christentum
Perspektiven für eine aufgeklärte Theologie
Mit David Steindl-Rast
Frankfurt/M.: Fischer Verlag, 2015

Le temps du changement
Science, société et nouvelle culture
Paris: Rocher, 1994

Das Tao der Physik
Die Konvergenz von westlicher Wissenschaft und östlicher Philosophie
Neue und erweiterte Auflage
München: O.W. Barth bei Scherz, 2000

Ursprünglich erschienen 1975 bei Droemersche Verlagsanstalt in Hamburg

Le tao de la physique
Paris: Sand & Tchou, 1994

Lebensnetz
Ein neues Verständnis der lebendigen Welt
München: Scherz Verlag, 1999

Verborgene Zusammenhänge
München: Scherz Verlag, 2002

Personal Notes

www.ingramcontent.com/pod-product-compliance
Lightning Source LLC
Chambersburg PA
CBHW020725180526
45163CB00001B/104